Lenilda Teixeira da Silva
Pedro Cavalcanti
Naisandra Bezerra

Morphology of the Digestive Tract of the Fish Steindachnerina notonota

Lenilda Teixeira da Silva
Pedro Cavalcanti
Naisandra Bezerra

Morphology of the Digestive Tract of the Fish Steindachnerina notonota

Anatomical and Histological Aspects

ScienciaScripts

Imprint
Any brand names and product names mentioned in this book are subject to trademark, brand or patent protection and are trademarks or registered trademarks of their respective holders. The use of brand names, product names, common names, trade names, product descriptions etc. even without a particular marking in this work is in no way to be construed to mean that such names may be regarded as unrestricted in respect of trademark and brand protection legislation and could thus be used by anyone.

Cover image: www.ingimage.com

This book is a translation from the original published under ISBN 978-3-330-77241-0.

Publisher:
Sciencia Scripts
is a trademark of
Dodo Books Indian Ocean Ltd. and OmniScriptum S.R.L publishing group

120 High Road, East Finchley, London, N2 9ED, United Kingdom
Str. Armeneasca 28/1, office 1, Chisinau MD-2012, Republic of Moldova, Europe
Managing Directors: Ieva Konstantinova, Victoria Ursu
info@omniscriptum.com

Printed at: see last page
ISBN: 978-620-8-64270-9

CONTENTS

I dedicate all the characters in this work to my Triune God. He is the True Principal Author!

ACKNOWLEDGMENTS

To the Sacred Heart of Jesus, thank you for your immense love and for always being with me, being faithful and the "strength of my life".

To the Immaculate Heart of Mary, my protection and sublime love.

To my parents, who are my "safe haven", my greatest encouragers, I thank you for your understanding and love.

To my sisters and my family, for being part of my life and for their support.

To my friends, for their encouragement and help.

To Prof. Dr. Naisandra Bezerra da Silva Farias, thank you for your kindness in accepting me as your advisor, for your patience and support in carrying out this work. A competent histologist! Thank you very much!

To all the professors of the Department of Morphology and the Master's Program - PPGBIOEF, for having contributed to my education by providing the basis for instigating morphological knowledge during my academic career at UFRN.

To the fishermen, for collecting the specimens and for sharing their popular knowledge, so important for the research.

To my fellow master's students, for their friendly words and help at various times.

To some of the authors cited in the references, for their kindness in sending their work.

To the technical support of Maria de Lourdes Freitas, Melyna Soares de Souto, and Maria Sara Maia de Queiroz, from the Laboratory of Histological Techniques, Department of Morphology - DMOR (Biosciences Center/UFRN), for their contribution to histological processing.

To the technical support of the Materials Structural Characterization Laboratory, of the

Department of Materials Engineering - DEMAT (Technology Center/UFRN), for the microphotographs obtained by Scanning Electron Microscopy.

For the acquisition and analysis of microscopic images carried out at the Edmond and Lily Safra International Neuroscience Institute (IIN-ELS), in Macaiba-RN, Brazil, with a microscopy system funded by FINEP; with the indispensable technical support of Pedro de França Cavalcanti, and for the receptiveness of the Institute's coordination.

Whatever you do in word or deed, do it in the name of the Lord Jesus Christ. Through him give thanks to God the Father.

Whatever you do, do it from the heart, as for the Lord and not for men.

(Col 3:17, 23-24)

SUMMARY

SILVA, L. T. **Morphological adaptations of the digestive tract of the neotropical fish *Steindachnerina notonota* (Characiformes, Curimatidae) to the detritivorous feeding habit.** 2016. 116 f. Defense (Master's Degree in Structural and Functional Biology) - Center for Biosciences, Federal University of Rio Grande do Norte, Natal, 2016.

Detritivory is common in dulcicolous fish, especially in the Neotropics, and for many species information on the digestive tract is scarce, as is the case with *Steindachnerina notonota*, which is ecologically important as a link in the food chain, serving as food for carnivorous fish and acting to purify aquatic ecosystems subject to organic pollution. The aim of this study was to analyze the morphological adaptations of the digestive tract of this endemic species of the caatinga, a region of the Brazilian semi-arid region. Thus, 10 necropsied adult specimens (males and females) were used for macroscopic and microscopic analysis of the digestive tract. All the structures were observed under a stereomicroscope. The organs were fixed in 10% formaldehyde and subjected to Hematoxylin and Eosin staining and PAS - Periodic Acid Schiff staining (gill tracks, epibranchial organ, esophagus, stomach, middle intestine and posterior intestine) and analyzed under light microscopy. The gill tracks were also fixed in 10% paraformaldehyde for analysis by scanning electron microscopy. The specimens analyzed ranged in total length from 6.4 to 13 cm, standard length from 4.8 to 10.3 cm, and weight from 3.23 to 35.92 g. The structural adaptations in the digestive tract were: a terminal type mouth; a particular buccopharyngeal complex; the absence of teeth and a tongue; three types of gill track shapes with mucous cells and taste buds; an epibranchial organ with its own peculiarities; a mechanical stomach with highly developed musculature in the pyloric region; and a very long and coiled intestine, which has an exclusive helical fold on the inside. The morphological aspects of *S. notonota*'s digestive tract related to its specialized detritivorous-iliophagous feeding habits.

Keywords: Morphology. Digestive tract. Detritivory. Iliophagous fish. *Steindachnerina notonota*.

1 INTRODUCTION

1.1 General Considerations about Fish

Fish are the group with the greatest diversity among vertebrates, accounting for more than half of all vertebrate species, and they show a great deal of known genetic variation; comprising around 33. 200 species, they show great adaptive success in marine and freshwater ecosystems.200 species, they exhibit extensive adaptive success in marine and freshwater ecosystems; and due to the high rate of descriptions of new species, this number has been growing steadily (TORRES; MATOSO; ARTONI, 2004; BRAGA; BORNATOWSKI; VITULE, 2012; MALABARBA; MALABARBA, 2014; FROESE; PAULY, 2016).

The Neotropical Region is known as one of the six largest zoogeographic regions on the planet, its ichthyofauna has a high degree of endemism, and it also has the greatest biodiversity of dulcicolous fish in the world (MALABARBA; MALABARBA, 2007; CHELLAPPA et al. 2009).

The global diversity of freshwater fishes has so far been estimated at almost 13,000 species; with the dulcicolous ichthyofauna of the Neotropical region constituting the most diverse in terms of number of species and population density, it contributes almost 31% of the world's fish population. And possibly around 30 to 40% of the freshwater fish fauna of South America has yet to be described; and many of the species discovered still have no studies reporting on their biology (VARI; MALABARBA, 1998; TORRES; MATOSO; ARTONI, 2004; AGOSTINHO; PELICICE; GOMES, 2008; LÉVÊQUE et al., 2008; CHELLAPPA et al., 2011).

Brazil has the greatest biological diversity of the seventeen megadiverse countries on the planet, with around 13% of the world's total biota. In this context, the country ranks first in the world in the number of freshwater fish, i.e. the most diverse dulcicola ichthyofauna, representing around 55% of species; It is also home to the Caatinga biome, which contains 57% endemism in fish, and is the only biome restricted to Brazilian territory, covering 70% of the semi-arid climate region of northeastern Brazil, with part of it in the state of Minas Gerais (ALVIM; PERET, 2004; TORRES; MATOSO; ARTONI, 2004; LEAL et al., 2005; QUEIROZ; RAPINI; GIULIETTI, 2006; BUCKUP; MENEZES; GHAZZI, 2007; CECAT, 2016).

The freshwater ictic species that occur in the Caatinga represent a fauna that is very well adapted to the climatic conditions and unstable hydrological regime of the region, with only the most resistant species surviving; in addition to suffering the influence of anthropogenic action, which has had a serious impact on reducing the composition of the native fauna (ROSA et al., 2003; CHELLAPPA, 2011).

With regard to the composition of the ichthyan fauna of the river basins of Rio Grande do Norte, among the seven most important (Piranhas-Assu, Apodi-Mossoró, Trairi, Ceará-Mirim, Potengi, Jacu and Curimataù), a 2014 study found that 64% of the species are native and 6% are exotic, with the order Characiformes being the most representative with 42% of species (NASCIMENTO et al., 2014). Recently, there was a report of a new species, the rattlesnake *Parotocinclus seridoenses*, found in the semi-arid region of Rio Grande do Norte, in the region of the upper Piranhas-Assu river basin, which is probably endemic to the semi-arid Caatinga (RAMOS et al., 2013).

Fish are extremely important in many ways, for example for ecological balance, socio-economic factors, academic and scientific knowledge and biomedical research.

Fish play a fundamental role in the food chain of aquatic ecosystems, and are an important nutritional resource for human beings, accounting for 16.7% of the world population's consumption as a source of animal protein, which was recorded in 2010 (FAO, 2014); fish activities generate billions of dollars in direct and indirect income worldwide, being significant in the economy of several countries (BRAGA; BORNATOWSKI; VITULE, 2012).

Knowledge of the local fish fauna contributes to future sustainable fisheries management plans, as well as the maintenance of native species and those with high commercial value; for this reason, inventories of the ichthyofauna are also relevant to understanding fish diversity and conservation, especially considering the speed with which human actions affect aquatic ecosystems (SAMPAIO et al., 2012).

In the field of teaching and research, fish are involved in studies of evolutionary biology, ecomorphology (TEIXEIRA; BENNEMANN, 2007; SAMPAIO; GOULART, 2011; SILVA-CAMACHO et al., 2014), ethnoecology (RAMIRES et al. 2015), animal behavior, and fish farming; the latter currently in Brazil using biotechnological methods, such as the genetic improvement of some species (BLANCK; TAKAKO, 2015).

The dulcicola fish known as Tilapia, is being applied in research as a biomaterial in regenerative medicine, due to the potential and biological safety of the collagen obtained from scales, skin and bone, as this collagen has bioactive properties, including excellent biocompatibility, low antigenicity and high biodegradability and cell growth potential (YAMADA et al., 2014; YAMAMOTO et al., 2014).

The species *Danio rerio* - zebrafish - is a freshwater fish that was initially used as a good *in vivo* model organism for developmental biology, in various studies of embryogenesis, cell migration, organ formation, as well as behavior, sleep, neuroscience and aging. Currently, its use has expanded into other areas of biomedical research, such as pharmacology, clinical research as a model for various diseases (e.g. cardiovascular and cancer), and dermatological research into the healing of skin wounds (CHAKRABORTY et al., 2009; RAKERS et al., 2010; KARDASH, 2012; RICHARDSON et al., 2013; HUITING; LAROCHE; FENG, 2015).

Recent studies are elucidating the degree of conservation between humans and zebrafish, which indicate the presence of many conserved genes, for example because they share 82% of disease-associated targets and a large number of drug metabolism pathways; in addition, the physiology of this fish is often better conserved among humans than rodent physiology (for example, in cardiac electrophysiology); due to its wide application, this species has become a favorite and promising model in biomedical research (KARDASH, 2012; MACRAE; PETERSON; 2015).

Fish of the order Characiformes are widespread in the freshwaters of Africa and the Neotropical region; and in the latter, especially in the large river systems of South America, there is a high proportion of detritivorous species from two closely related families: Prochilodontidae and Curimatidae (BOWEN, 1983; GUISANDE et al., 2012).

Among the diversity of Neotropical fish, this study will focus on the teleost *Steindachnerina notonota* (Curimatidae), popularly known in Rio Grande do Norte as Piabussù. This species is endemic to the caatinga and is important for maintaining the ecological balance in the freshwater environments of the Brazilian semi-arid region, as it is a detritivorous animal. For this reason, it has a promising future as an environmental bioindicator, given that many pollutants accumulate at the bottom of water bodies. There is little information on its digestive tract, so many structural aspects that have not yet been described will be characterized. This study will focus on the morphological adaptations of its digestive tract and how they relate to its feeding habits.

2 LITERATURE REVIEW

2.1 Importance of the Morphological Study of the Digestive Tract

Fish show morphofunctional diversity in the digestive tract, especially teleosts, due to the fact that they have a large number of species and occupy different positions in the food chain. At the same time, the diversity in all the body planes of these animals is notable, both in the external and internal anatomy (MALABARBA; VARI, 1998; KHOJASTEH, 2012).

The history of research into the morphology of the digestive tract of fish began with the first simple description of their general anatomy, by Aristotle over 2000 years ago in his work *Historium Animalia.* With the advent of comparative anatomy studies, there was a significant effort to understand the anatomical diversity of fish and vertebrates in general, starting with the works of Cuvier - 1805, Rathke - 1820 and 1837, Edinger - 1877, Oppel - 1896, and Hopkins - 1895. Advances in technology, from optical microscopy, scanning and transmission electron microscopy; staining techniques, such as histochemistry and immunohistochemistry; and others, have contributed in such a way that the level of detail has progressed to the study of organs, tissues, cells, and even at the molecular level (WILSON; CASTRO, 2011).

The digestive tract of fish has attracted the interest of many researchers due to the high degree of structural variation, which is much greater than in other vertebrates, and due to the fact that ictic species show diverse feeding habits and behaviors (MENIN, 1988).

Morphofunctional studies of the digestive tract of fish are of great relevance because:

(1) They are considered effective tools for understanding the mechanisms of food intake, digestion and absorption, i.e. they provide a context for understanding the topographical organization of specialized organs; consequently, for an essential elucidation of the related anatomical and histophysiological mechanisms (KHOJASTEH, 2012; WILSON; CASTRO, 2011; SANTOS et al., 2015);

(2) They identify differences in the macro and microscopic anatomy of the alimentary canal between various species (KHOJASTEH, 2012), enabling a comparative and

evolutionary approach, both among fish and with other vertebrates;

(3) They make it possible to infer the feeding habits of different species with a good margin of certainty (MENIN, 1988), and can add analysis of stomach contents, thus clarifying aspects related to the characteristics of the feeding site, i.e. providing support for interpreting the feeding dynamics and spatial occupation of fish (feeding ecology/behavior) (SEIXAS FILHO et al., 2001; BECKER et al., 2010; PESSOA et al., 2013);

(4) They are related to the individual's stage of development (ANGELESCU; GNERI, 1949; SEIXAS FILHO et al, 2001), meaning that ontogenetic studies are possible;

(5) They act as a certainty in the systematic identification of species, for example, by the oral dentition, or by the arrangement of the intestinal loops (HOWES, 1982; HARDER, 1958; SILVA et al., 2004);

(6) They demonstrate the specialization of organs in performing other non-digestive functions, such as osmoregulation, respiration and defense (WILSON; CASTRO, 2011);

(7) They are related to nutritional status (ANGELESCU; GNERI, 1949; SEIXAS FILHO et al, 2001): They can be applied in nutritional development research (such as the improvement of fish nutrition protocols), in the preparation of diets and in food management, thus providing a basis for studies aimed at implementing techniques for the cultivation of species with economic value (ROTTA, 2003; HERNANDEZ; PÉREZ GIANESELLI; DOMITROVIC, 2009; KHOJASTEH, 2012);

(8) They clarify pathological and/or physiological changes, whether related to infectious diseases, environmental contamination or artificial diets (applied in fish farming); since fish can act as bioindicators of environmental pollution, for example, mainly serving as a basis for toxicological analysis, since intestinal areas absorb particles of contaminated food, as well as changes in the structure of gill tracks, resulting from the influence of pollution (RIBEIRO; FANTA, 2000; CARRASSÓN et al., 2006);

(9) They contribute to the use of biological information for the management of species (SANTOS et al., 2015), especially in programs to preserve native fish for their sustainable use, so as not to trigger imbalance in the ecosystem or extinction;

(10) In physiological research: such as the identification of genes that mediate vertebrate digestive physiology, for example, the Zebrafish is a valuable model for mammalian lipid metabolism (FARBER et al., 2001; ANDERSON; CARTEN; FARBER, 2011);

(11) And because such morphological studies are still scarce for Neotropical ichthyofauna (CANAN et al., 2012; PESSOA et al., 2013).

2.2 Structural aspects

The digestive tract of fish is made up of the mouth, the buccopharyngeal cavity and the digestive tract, which generally consists of the esophagus, stomach and intestine, as well as pyloric cecums (or intestinal cecums) (ZAVALA-CAMIN, 1996).

Topographically, the most widely used division in Brazil, the digestive tract of fish is divided into four regions: cephalic intestine (mouth and buccopharyngeal cavity), anterior intestine (esophagus and stomach), middle intestine (intestine proper) and posterior intestine (final portion of the tract, and which includes the rectum); However, in some cases, there is no clear morphological distinction between the midgut and the hindgut; and embryologically, the epithelium of the foregut is of ectodermal origin, and that of the midgut is of endodermal origin (BÉRTIN, 1958; MENIN, 1988; CASTRO, 2002; WILSON; CASTRO, 2011). The aforementioned structures making up the tract exhibit many variations, as well as folds, valves, sphincters and many other particularities.

The digestive system is made up of the digestive tract and the attached glands, which are the gallbladder, pancreas and liver; in the case of some species, the pancreatic tissue is located diffusely in the liver, called intrahepatic pancreas or hepatopancreas; and fish do not have salivary glands (ROTTA, 2003).

The digestive tract is shared with the respiratory system, mainly due to the presence of gills and oral valves (these are involved in respiratory mechanics) (RODRIGUES; MENIN, 2006b) in the buccopharyngeal cavity, and in some cases, due to the presence of a connection between the bladder and the esophagus through the pneumatic duct, in the so-called physostome fish.

Há the so-called accessory air-breathing organs of the digestive tract, which evolved in a number of fishes as a means of surviving periods of aquatic hypoxia; these organs are: the esophagus (families Umbridae and Blenniidae), the stomach (Loricariidae and Trichomycteridae) and the intestine (Cobitidae and Callichthyidae), which have morphological characteristics that are indicative of increased gas exchange, including: increased vascularization; presence of intraepithelial capillaries with a concomitant reduction in the blood-lumen distance; increased luminal surface area; presence of surfactant-producing Iamellar bodies; and presence of muscular sphincters isolating the respiratory region from other regions of the tract (WILSON; CASTRO, 2011; NELSON, 2014).

In most of the fish species investigated, the digestive tract is structurally well innervated, for example, the mouth and perioral region are rich in nerve endings (ROTTA, 2003); as well as the presence of autonomic innervation which includes both local intrinsic nerves (the enteric nervous system) and extrinsic nerves (sympathetic and parasympathetic) (OLSSON, 2011).

2.3 Eating habits

Fish need nutrients from food for growth, reproduction and survival (LAGLER, 1977). According to Nikolsky cited by Moraes, Barbola and Guedes (1997), ichthyan species are adapted to feeding on a particular item or set of items (wide variability).

In most teleosts, trophic plasticity occurs, which is an especially striking aspect attributed to tropical dulcicolous ichthyofauna, in which fish can switch from one food to another, due to the preference of a particular food item in a given period, or if there are changes in the relative abundance of certain food resources. This trophic flexibility is inherent to spatial, temporal, ontogenetic, individual and behavioral variations (ABELHA; AGOSTINHO; GOULART, 2001). According to Fugi and Hahn (1991), "knowledge of the feeding habits of fish is of fundamental importance in delineating the trophic structure of the ecosystem, as well as the trophic level occupied by the species".

In the literature, there are many classifications of trophic categories in fish, and according to Wilson and Castro (2011), fish are classified into the following feeding habits, based on the nature of the food: herbivores, carnivores, omnivores and detritivores. However, the designations of subdivisions of these categories are broad, and sometimes problematic, because the authors make their own designations/subclassifications.

Wootton, quoted by Andrian, Peretti and Bosco (2005), classifies herbivores into: browsers - they feed on parts of higher plants, grazers - they exploit algae from substrates and phytoplanktivores.

Species that feed on live animals are generalized as carnivores, and the specialized subdivisions of this category are varied, depending on the author's denomination (ANDRIAN; PERETTI; BOSCO, 2005); for example, they can be: ichthyophagous or piscivorous (fish), carcinophagous (shrimp), malacophagous (molluscs), teutophagous (cephalopods), insectivorous (insects), invertivorous (invertebrates), hematophagous (blood and mucus) and lepidophagous (scales of other fish). Several authors use other complementary terms, for example, microphages and macrophages (STEVENS; HUME, 2004), as well as filter feeders, planctivores or plankton feeders, benthophages, scavengers, frugivores and granivores (HORN et al. 2011).

Omnivorous fish have a mixed diet, often according to the availability of the environment; thus, the authors explain about omnivory by often considering the main foods, for example, they cite as: omnivorous with a strong tendency towards carnivory, a tendency towards herbivory, a tendency towards insectivory, or a tendency towards frugivory.

2.4 Trophic Morphological Adaptations

According to Pillay quoted by Moraes, Barbola and Guedes (1997), the nature of the food ingested will depend firstly on morphology and feeding behavior, and secondly on the quantity and composition of the food available in the environment. According to Fugi (1993), although there is considerable plasticity in the diet of fish from tropical communities, the morphology of the digestive tract of many species restricts the food spectrum.

The morphology of the gastrointestinal tract of fish shows many specific variations related to diet, phylogeny of feeding habits, body shape, as well as characteristics that reflect functional differentiation (BECKER et al., 2010).

Several authors have reported a correlation between the morphology of the digestive tract and the feeding habits of fish, i.e. differences in anatomical and histological structural characteristics make it possible to diagnose the feeding regime of a species, whose characteristics are identified as adaptations and modifications (SEIXAS FILHO et al., 2001; SILVA, GURGEL and SANTANA, 2005).

Adaptations are morphological variations resulting from the action of environmental factors on the organism, of a permanent nature, produced in phylogenetic evolution; while modifications are morphological variations of a temporary nature, produced in the ontogenetic cycle of the individual (ANGELESCU; GNERI, 1949; SEIXAS FILHO et al., 2001). It should be noted that some adaptations in fish are possibly non-existent compared to terrestrial vertebrates, since certain foods available are only found in the aquatic environment (ROTTA, 2003).

The trophic morphological adaptations of fish can be evidenced by the presence, position, shape and size of a particular structure or organ and its various peculiarities (ROTTA, 2003).

The mouth and buccopharyngeal cavity have great significance in inferences about eating habits related to the way food is selected, apprehended and conveyed, and the adaptations that can be analyzed are, for example: shape, position and size of the mouth; width of the oral cleft; shape of the oral rima; presence of barbels (chemoreceptive appendages - they have gustatory muscles, acting as taste receptors at a distance); structure of the lips; tongue; presence of oral valves (helps prevent food escaping); presence/absence of oral and pharyngeal teeth (type, arrangement and orientation of functional oral teeth; and pharyngeal teeth situated on the pharyngeal plates, as well as the presence of pharyngeal dentigerous areas); the presence of gill tracks; the shape, size and pattern of the mucosa lining the buccopharyngeal cavity; and the presence of the palatal or epibranchial organ (RODRIGUES; NAVARRO; MENIN, 2006; RODRIGUES; MENIN, 2006a; RODRIGUES; MENIN, 2006b).

Generally speaking, fish that feed on what is above them have their mouths in the upper position and live on or near the surface, or on the bottom; those with their mouths in the terminal position can explore in relation to taking food; and those with their mouths in the lower position feed on what is below them, near the bottom. And certain species have a protruding mouth, in which they protrude the bones of the maxilla and/or mandible, causing strong suction, favoring the capture of prey from a distance (ROTTA, 2003; HELFMAN et al., 2009; BEMVENUTI; FISCHER, 2010).

The lips are commonly fleshy in herbivorous fish, attributed to grazing and filtering; in carnivorous fish, they are generally thin and with few modifications; and some species have lips with taste buds, such as the dark rattlesnake (*Hypostomus commersoni*) (ROTTA, 2003), or very developed lips (Blennidae and Labridae), and others may be absent (Scaridae) (MAKINO, 2010).

Bony fish can exhibit teeth located on the lips, on the maxilla (maxillary and premaxillary), on the mandible (dentary), or located on the palate of the oral cavity (vomer, palatine and ectopterygoid), on the floor of the cavity (tongue) and in the pharynx; that is, also as pharyngeal teeth, for example, which can occur in cushions close to the gill arches, and form a single set (Figure 1) (LAGLER et al., 1977; ZAVALA-CAMIN, 1996; STEVENS; HUME, 2004; BEMVENUTI; FISCHER, 2010).

The teeth on the lips of Mugilidae fish are used to scrape off algae attached to hard substrates; in *Semaprochilodus* species, the teeth on the lips are used to remove food attached to trees, in this case, when these are submerged during floods (ZAVALA-CAMIN, 1996).

Fish have a wide variety of teeth, with different shapes and sizes (ROTTA, 2003). The types of teeth found in neotropical fish of the Characiformes order are illustrated in figure 2.

Generally, differences in dentition indicate differences in food preference and thus serve as a tool for assessing diet (ROTTA, 2003; SAMPAIO; GOULART, 2011).

Tooth morphology has been shown to be a significant phylogenetic indicator, as different types of teeth found in lineages of the order Characiformes, showed consistent trends for some types of teeth (or their absence) and a clear association with the types of food reported to be exploited by these species (Figure 3) (GUISANDE et al., 2012).

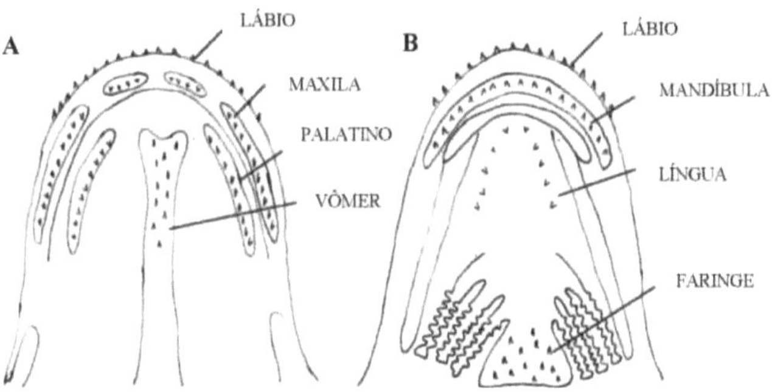

Figura 1 - Diagram of the buccopharyngeal cavity in horizontal section, showing the teeth in fish, in general - Adapted. **A** - Top of the cavity (maxilla). **B** - Floor of the cavity (mandible).

Source: SAMPAIO; GOULART, 2011.

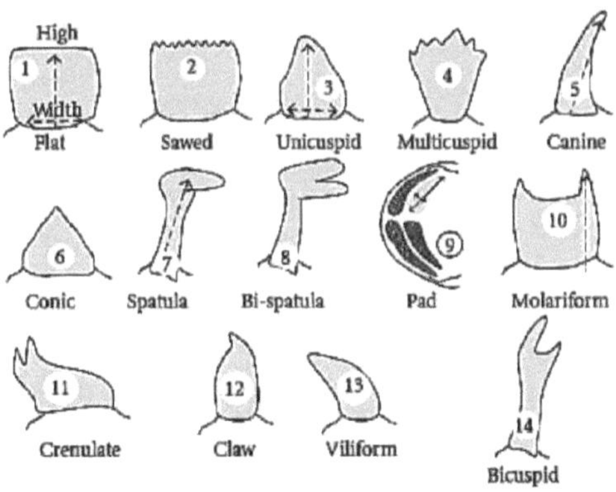

Figura 2 - Types of teeth according to shape found in neotropical fish of the order Characiformes.

Source: GUISANDE et al., 2012.

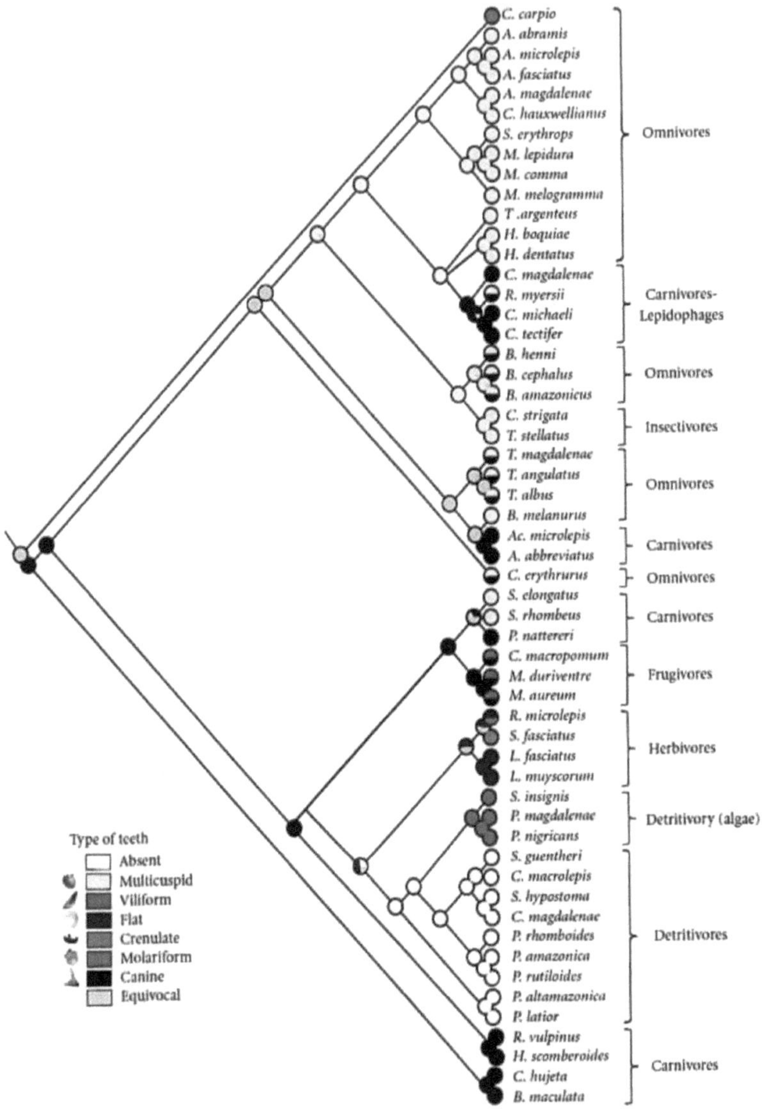

Figura 3 - Cladogram of the reconstruction of the types or absence of teeth correlated with the phylogeny of the feeding habits of neotropical fish of the order Characiformes.

Source: GUISANDE et al., 2012.

The oral teeth have the function of seizing prey, as well as helping to swallow and scrape food (RODRIGUES; MENIN, 2006a). In carnivorous, predatory species, they commonly have canine and pointed teeth; in herbivorous and omnivorous species, in contrast to carnivores, the teeth are not very developed, as this is correlated with the nature of the diet, being composed, for example, of small animals, seeds and algae (RODRIGUES; MENIN, 2005; RODRIGUES; MENIN, 2006b).

The pharyngeal teeth have been studied in correlation with possible functions according to different feeding habits: in carnivores, as a food press (prevents prey from escaping); in herbivores, the grinding of food; in omnivores, the maceration of soft-bodied organisms and algae (SCHUINGUES et al., 2013).

Gill trails are cartilaginous or bony formations, generally elongated, lined up on the inside of the gill arches, presenting adaptations related to the size of the food ingested, which involve the following functions: (a) to aggregate particulate food; (b) to prevent the escape of food material from the buccopharyngeal cavity (in some cases, the tracks have denticles, which promote the capture, seizure and even scraping of food); (c) to retain (by filtering) small foods that could escape through the tracks; (d) assisting in orientation during swallowing, for example, either to position larger food items before they go down the esophagus, or to prevent regurgitation; (e) as well as participating in the protection of gill filaments (ZAVALA-CAMIN, 1996; RODRIGUES; MENIN, 2006b; SAMPAIO; GOULART, 2011).

For example, carnivorous fish can have gill tracks with varying degrees of development, being more rudimentary in ichthyophagous fish; and in planktivorous fish, the tracks are thin, long, numerous and very close together (ANGELESCU; GNERI, 1949; RODRIGUES; MENIN, 2006a; SAMPAIO; GOULART, 2011).

Species of the Mugilidae family have secondary trails in the form of small hooks, emerging from the trail itself, which increases their filtering capacity; as with other fish, some species have mucus in their trails, which increases their retention power compared to smaller organisms (ZAVALA-CAMIN, 1996).

In the posterior region of the buccopharyngeal cavity, there are two types of structures (organs), called the palatal organ and the epibranchial organ.

The palatal organ is a muscular and chemosensitive structure that forms the roof of the anterior pharynx, responsible for the selective retention of food particles, and is present inside the buccopharyngeal cavity of cypriniform fish; because these fish lack the oral and palatal teeth, all food maceration takes place in the pharyngeal region; and this is especially important in fish that feed on benthic suction, such as the sucking fish of the Catostomidae family (most feed on benthic macroinvertebrates, plant material, algae, or zooplankton, but some species are pelagic planktivores), and some of the Cyprinidae family (Figures 4 and 5) (CALLAN; SANDERSON, 2003; FINGER, 2008; DOOSEY; BART JR, 2011; PEREIRA, 2014).

The morphology of the palatal organ varies considerably among cypriniform fish, being more diverse in terms of size and shape in the Catostomidae family. In the four species of the Characiformes order studied (*Astyanax mexicanus*, *Bryconamericus ricae*, *Curimata peruana*, and *Ichthyoelephas humeralis*), there is no palatal organ on the roof of the anterior pharynx (DOOSEY; BART JR, 2011).

The synonyms of terms and phrases used to describe the palatal organ in cypriniform fish were: "Geschmacksorgan"; "organe palatin"; "Palatal organ of Valatour"; "Palatal organ"; "Muscular cushion"; "Pharyngeal pad"; "Thick pad of connective tissue"; "Large, thick pad of connective tissue and epithelium"; "Supra- pharyngeal pad"; "Palatine organ"; "Pharyngeal cushion"; "Posterior papillose pad". However, the term palatal organ is chosen because it is among the oldest and most consistently applied names in the literature (DOOSEY; BART JR, 2011).

In the posterior part of the pharynx is a region occupied by a keratinized chewing pad (Figure 4), in which this structure is at the posterior end of the palatal organ, present in catostomid fish, or in a skeletal process in cyprinids; Its shape varies, but it is usually ovoid, or in cyprinids diamond-shaped, and in Catostomidae, the shape is semilunar and the lateral margins extend dorsally around the organ; and it serves as an occlusion surface for the pharyngeal teeth, as the pharyngeal teeth grind against the pad to macerate the food (DOOSEY; BART JR, 2011).

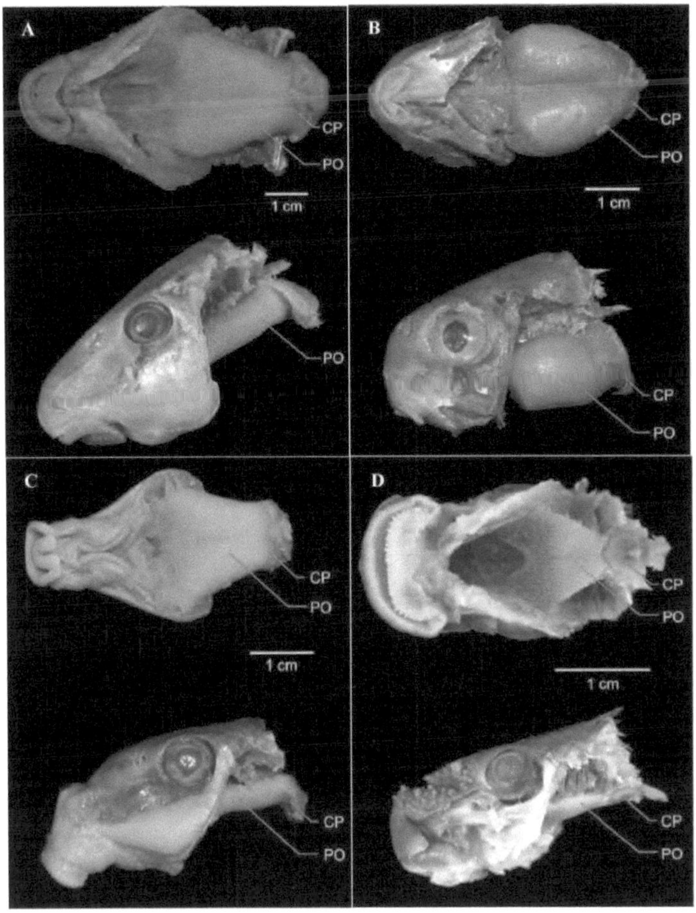

Figura 4 - Dorsal parts of the heads illustrating the palatal organ (PO) and chewing pad (CP) in ventral (top) and lateral (bottom) views. Opercula and pharyngeal (gill) arches were removed from the specimens. **A** - *Catostomus occidentals*. **B** - *Carpiodes velifer*. **C** - *Hypentelium etowanum*. **D** - *Labeo cylindricus*.

Source: DOOSEY; BART JR, 2011.

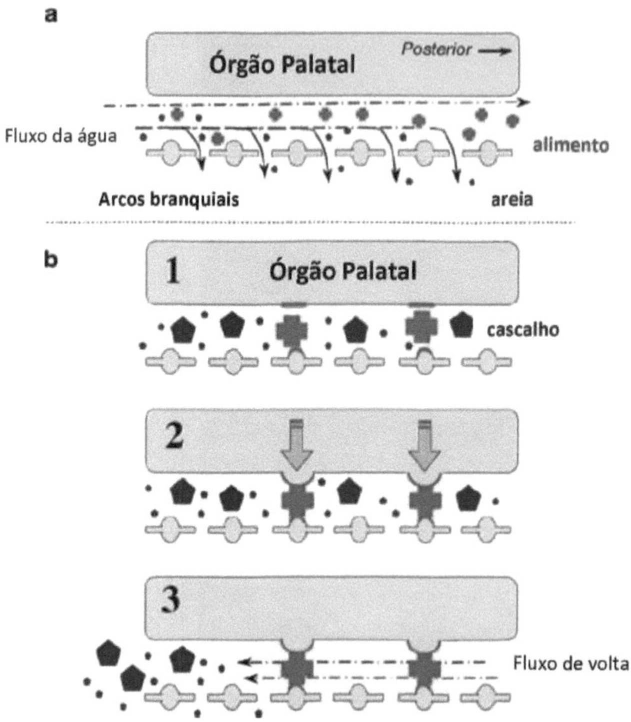

Figure 5 - Schematic drawings of how the palatal organ and gill apparatus separate food from the substrate under different conditions. **a** - When food particles are small enough to be suspended in water, the gill arches serve as a cross-flow filter (Callan; Sanderson, 2003). As the water passes through the gill arches, the heavier particles, e.g. sand; fall out while the lighter food (small red crosses) is carried backwards towards the posterior pharynx. **b** - A sequence (1-3), showing how the larger food particles and larger substrate particles (e.g. gravel); the food is actively sorted by trapping its particles (red cross) between the palatal organ and the floor of the mouth. **(1)** The food particles activate the gustatory muscles over the palatal organ and the gill arches (red lines). **(2)** This triggers the protrusion reflex of the palatal organ to press the food particles against the gill arches and hold them in place. **(3)** The flow back then results in the inorganic material being expelled out of the mouth or through the spaces between the arches.

Source: FINGER, 2008.

The epibranchial organs are enigmatic tubular structures, bilaterally paired, found in the buccopharyngeal cavity, exactly in the posterior portion of the pharynx, of many fish that feed by filtration (microphages); and assume at least seven different types of morphologies, in which they differ in position, shape and structure (Figures 6 and 7); being present in some unrelated families of fish (HANSEN et al., 2014; HOLLEY et al., 2015). However, both the families that have these organs and their morphological types are in need of an updated and consistent review.

In most cases, the epibranchial organs comprise a series of blind sacs or tubes, but sometimes, these tubes are open (continuous) or vestigial; in which, these tubes always align with the branchial arches and are supported by modified epibranchial bones; and in many cases, these organs are exclusively associated with the fourth branchial arch (right and left) (HANSEN et al., 2014).

The term "epibranchial organs" is more commonly used nowadays, but it should be noted that over the years, various authors have used many different names, such as "Kiemenschnecke"; "accessorischen kiemenorgane"; "epigonal kiemenorgane"; "diverticule pharyngien": "Kiemenschnecke"; "accessorischen kiemenorgane"; "epigonal kiemenorgane"; "diverticule pharyngien"; "diverticule pharyngien suprabranchial"; "sac pharyngien"; "organo epibranchiale"; "accessory branchial organ"; "organes pharyngiens accessoires"; "organe suprabranchial"; "gillhelix"; "pharyngeal pouch"; "rachensacke"; "pharyngeal pockets", "pharyngeal organs"; "suprabranchial organ", "gill snail"; "accessory gill organ"; "crumenal organ", and others (NELSON, 1967; BERTMAR; KAPOOR; MILLER, 1969; MENIN, 1988; MENIN; MIMURA, 1991; PASLEAU; DIOGO; CHARDON, 2009; HANSEN et al., 2014).

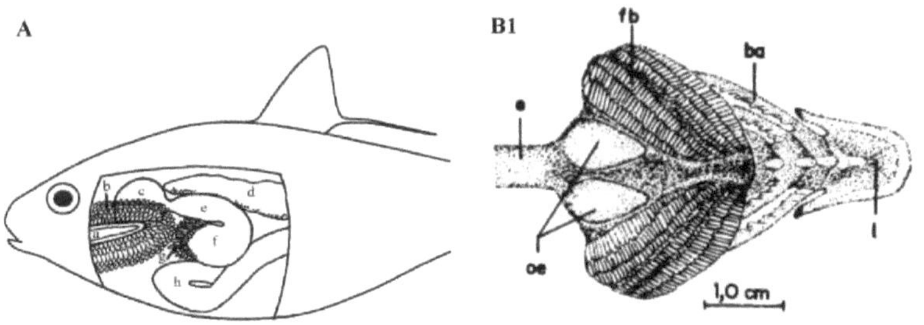

Posição dos órgãos epibranquiais de *Prochilodus affinis* (21,5cm, F). Vista dorsal do intestino cefálico: ba - bordo anterior dos arcos branquiais; e - esôfago; fb - filamentos branquiais; l - língua; oe - órgãos epibranquiais.

B2

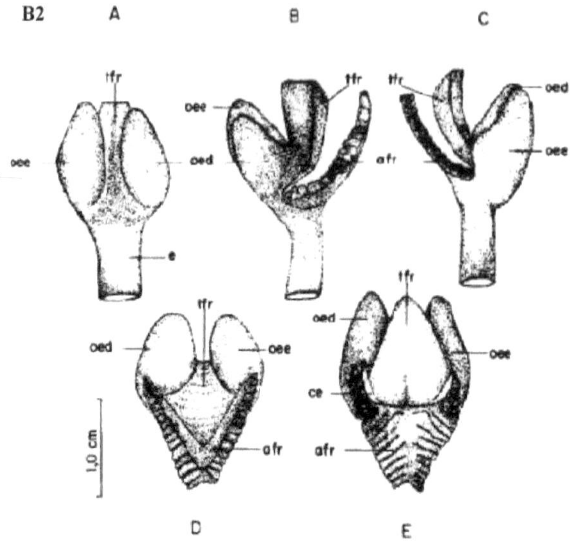

Órgãos epibranquiais de *Prochilodus affinis* (21,1cm, F). A. Vista póstero-dorsal; B. Vista lateral direita. C. Vista lateral esquerda; D. Vista frontal; E. Vista frontal, com o teto da faringe rebatido, evidenciando o canal de entrada.
afr - assoalho da faringe; ce - canal de entrada; e - esôfago; oed - órgão epibranquial direito; oee - órgão epibranquial esquerdo; tfr - teto da faringe.

Figure 6 - Epibranchial organs. **A** - Lateral view of the epibranchial organ of a specimen from the Clupeidae family. The *Dorosoma cepedianum* specimen with part of the body wall and operculum removed, to illustrate the locations of the structures: (a) buccopharyngeal cavity; (b) gill filaments on the gill arches; (c) epibranchial organ; (d) swim bladder; (e) oesophagus; (f) mechanical stomach (gizzard); (e+ f) anterior intestine; (g) pyloric ceca; (h) intestine. **B1 and B2** - Epibranchial organ of a specimen of the family Prochilodontidae; in *Prochilodus* they are type 1 (expanded primitive sac).

Source: MENIN; MIMURA, 1991; HOLLEY et al., 2015.

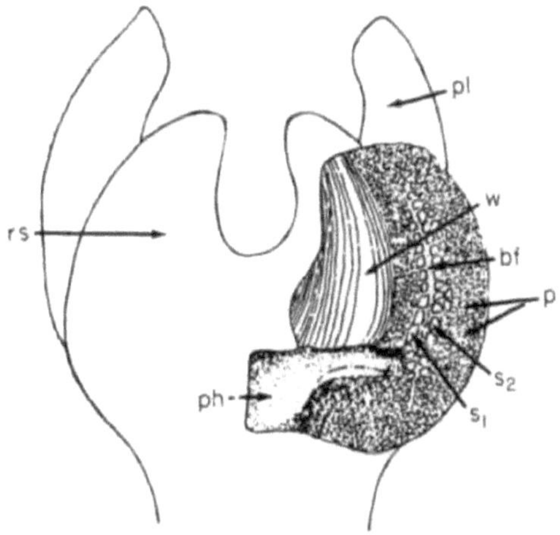

A Dorsal aspect of sac-like epibranchial organs (Type 1) of *Curimatus vittatus;* right sac opened. pl, fourth epibranchial; w, adipose tissue; bf, tributary canal; p, papillae; s_2, s_1, modified gillrakers; ph, pharynx; rs, epibranchial organ. (From Heim, 1935.)

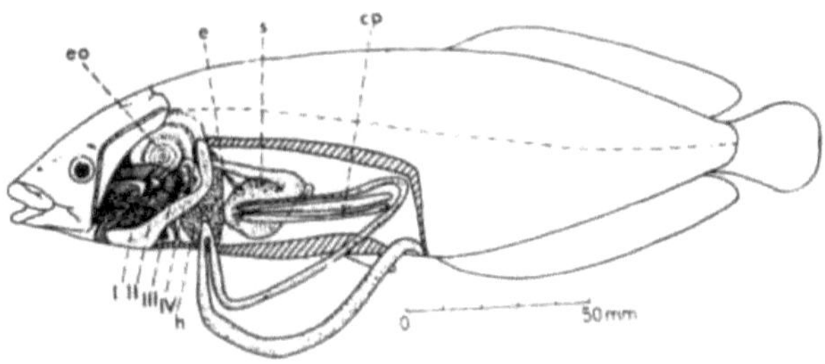

B1 Lateral view of a dissected osteoglossid, *Clupisudis niloticus,* showing position of spiral tube epibranchial organ (Type 3). eo, epibranchial organ; e, esophagus; s, stomach; cp, pyloric caeca; h, heart; I–IV, branchial arches. (Modified from d'Aubenton, 1955.)

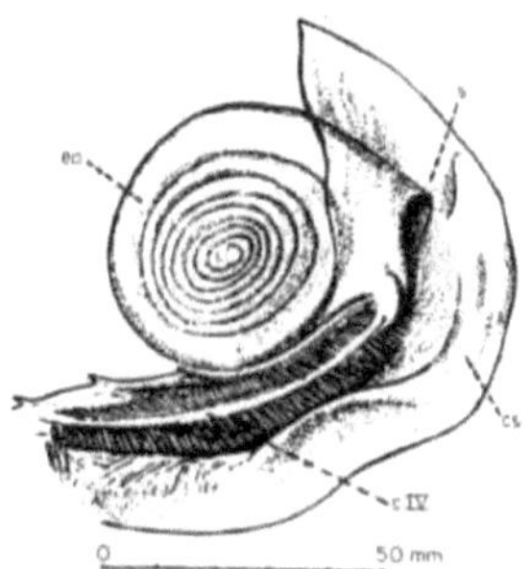

B2 Lateral view of epibranchial organ (Type 3) of *Clupisudis niloticus;* continuous tube extending in spiral with gillrakers present to terminus. eo, epibranchial organ; s, exit for water from epibranchial organ; cs, shoulder girdle; cIV, fourth ceratobranchial. (Modified from d'Aubenton, 1955.)

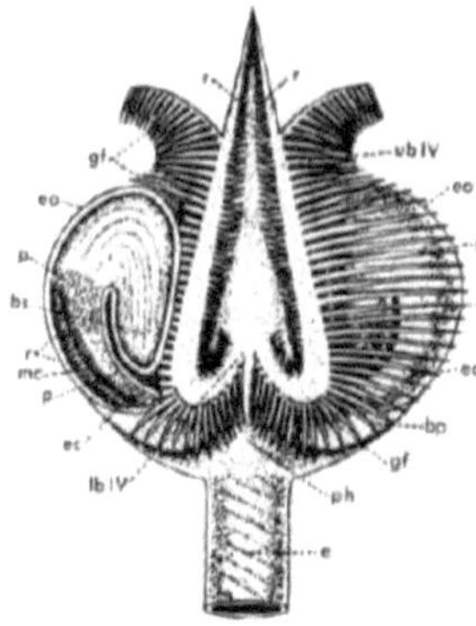

C

Diagram of ventral view of entrance canal-blind sac epibranchial organs of *Chanos chanos* (Type 5); right organ cut horizontally to show position of marginal channel and gillrakers along the entrance canal. From top left to top right: r, gill-rakers; gf, gill filaments; eo, epibranchial organ; p, papillae; bs, blind sac; me, marginal channel; ec, entrance canal; lbIV, lower branch of arch IV; e, esophagus; ph, pharynx; bp, bony plate; ubIV, upper branch of arch IV. (Modified from Kapoor, 1954.)

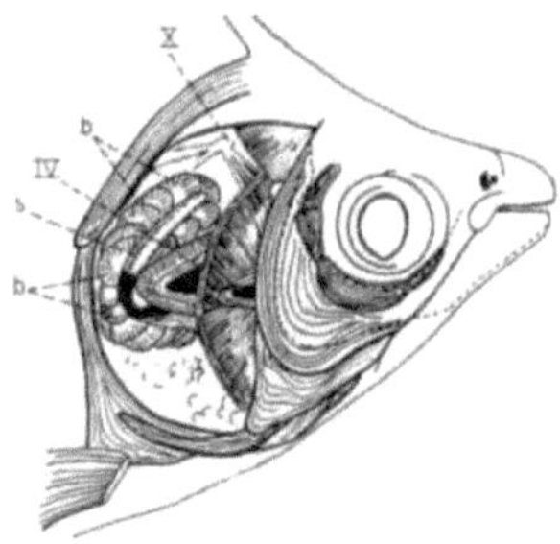

D

Dissected head of *Citharinus citharus* showing position of the tube and leaf-like lobuli that compose the epibranchial organ (Type 7). b, leaf-like lobuli; s, tube; IV, fourth branchial arch; X, vagus nerve. (Modified from Sagemehl, 1887.)

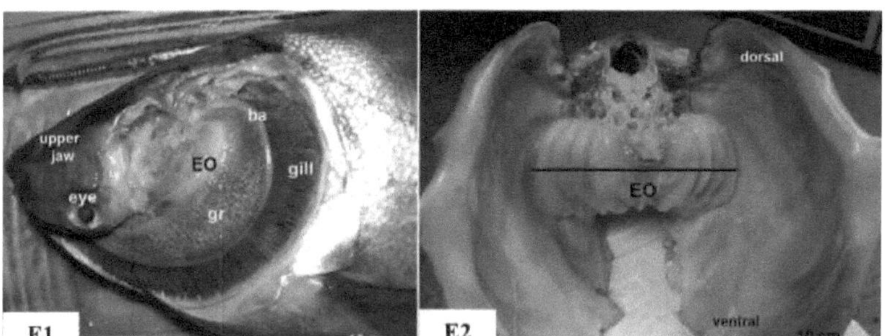

Gross morphology of the epibranchial organ in *H. molitrix* and *H. nobilis*. (E1) The epibranchial organ (EO) of *H. molitrix* is located behind the eye adjacent to the upper jaw and surrounded by the branchial arches (ba). The gill rakers (gr) are closely associated with the EO. (E2) Upper portion of the body of *H. molitrix* dissected to show paired EOs with four ridges on each side of the paired organs.

Figure 7 - Images of some morphological types of epibranchial organs. **A** - Type 1. **B1 and B2** - Type 3. **C** - Type 5. **D** - Type 7. **El and E2** - Type reported in 2014.

Source: BERTMAR; KAPOOR, 1969; HANSEN et al., 2014.

In teleosts, the esophagus is a tube, which can be short, wide or straight; and the stomach has various shapes and regions; and both organs are structurally adapted to the types of diet (FRANCO, 1994; MAKINO, 2010).

In piscivores, the stomach is quite elongated; in omnivores, it is a siphonoid sac and, in detritivorous-iliophagous fish (they ingest silt), there can be a stomach with a chemical region, and another mechanical region (the "gizzard" - analogous to that of birds); This is a highly muscularized structure without glands, whose function is to grind the ingested food together with the sediment of the substrate (SILVA, 2004; MÔNACO; RESENDE, 2012).

There are species that are called gastric and agastric (absence of a true stomach), and these can have the proximal part of the intestine enlarged, being a pseudo-stomach, called an intestinal bulb; with the temporary purpose of storing food, which may or may not be histologically differentiated from the true intestine, and recognized visually only with content (ZAVALA-CAMIN, 1996; OLSSON, 2009).

Agastric fish are, for example: Cyprinidae, Cobitidae, Cypridontidae, Gobiesocidae, Gobioidei, Blennidae, Scaridae (WILSON; CASTRO, 2011).

The intestine varies in shape, arrangement and length. According to Fryer and Iles quoted by Giora and Fialho (2003), the length of the intestine is directly related to the trophic category of the species, and is ordered as follows: carnivores< omnivores< herbivores< detritivores.

Blazquez (1995) points out that morphological variations and specializations relevant to feeding are considered a way of reducing competition between species, which is fundamental for maintaining the trophic chain in aquatic ecosystems.

2.5 Detritivorous fish

- **Detritivore and Iliophage**

Detritivorous fish consume particulate organic material deposited on the bottom or in submerged vegetation, which is often mixed with mineral particles with no nutritional value (inorganic sediments). This particulate organic material is derived from the bodies of dead organisms or fragments thereof, excretions left by living organisms, dead plants (rich in lignin and cellulose), small quantities of microinvertebrates, filamentous algae, as well as associated microorganisms such as fungi and bacteria (YOSSA; ARAÙJO-LIMA, 1998; PEREIRA, RESENDE, 1998).

According to Fugi (1993), detritivorous fish feed basically on organic detritus in moderate stages of decomposition (ingesting larger particles).

Within the detritivorous fish, there is a specialized diet, called the Iliophagus type[1] ; they consume almost exclusively fine particles of organic matter (in advanced stages of decomposition) from the mud, i.e. typically associated with soft sediments in lentic and lotic environments; which contains inorganic sediments (the most common part of the diet), organic detritus, algae, diatoms and periphyton (FUGI, 1993; MORAES; BARBOLA; GUEDES, 1997; BOWEN; GU; HUANH, 2006).

Detritivory is present in only a small percentage of fish species; and many species are occasional consumers (for example, they may switch to consuming detritus as a temporary diet when preferred foods are not available), while others are generalists and highly specialized (BOWEN, 1983; YOSSA; ARAÙJO-LIMA, 1998).

The best known dulcicolous detritivorous families are Prochilodontidae, Curimatidae (these two are typically detritivorous) and Loricariidae from the Neotropical region; some Citharinidae, Cichlidae and a few Cyprinidae; in addition to the Mugilidae Family, which has a wide geographic distribution, present in tropical and subtropical ecosystems all over the world, in coastal marine waters, estuaries and freshwaters, with most species typically estuarine-coastal (BOWEN, 1983; GERKING, 1994; DELARIVA; AGOSTINHO, 2001; BOWEN; GU; HUANH, 2006; FISCHER; PEREIRA; VIEIRA, 2011).

Fish from the Citharinidae family are native to Africa (BOWEN, 1983; MOREIRA, 2007).

---

1 This term has also been written in English as iliophago, illiophago, and ileophago, but the original spelling of ilyophago proposed by Allen in 1936 is the preferred one (BOWEN; GU; HUANH, 2006).

In the Cichlidae family (order Cichliformes), the geographical distribution covers Africa, the Jordan River valley in the Middle East, southern India and Sri Lanka, Madagascar, Iran, Cuba and Santo Domingo Island, Central America, North America and South America (FAVERO; POMPEU; VALLADARES, 2010).

In the Cichlidae family, for example, species of the *Oreochromis* and *Tilapia* genera, which are originally members of the African cichlids, have been introduced into various Brazilian lakes and reservoirs and are considered exotic species in the Northeast of Brazil (CHELLAPPA et al., 2003; LEÂO et al., 2011; ZAGANINI; VIDOTTO-MAGNONI; CARVALHO, 2012).

The Cyprinidae family (order Cypriniformes) is completely absent in the Neotropics, occurring in North America, Africa and Europe (BOWEN, 1983; BOWEN; GU; HUANH, 2006).

▪ Eating habits

Dulcicolous teleosts show trophic plasticity or adaptability (ABELHA; AGOSTINHO; GOULART, 2001), but this adaptation is limited by their feeding behavior, morphological apparatus and digestive capacity. According to Gerking (1994), in relation to their feeding habits, fish can be classified as:

- **Generalists** - do not show a marked preference for a specific food source, using a diet with a broad spectrum of foods (within their determined eating habits - personal observation);
- **Specialists** - imply a diet restricted to a relatively small number of items and usually characterized by remarkable morphological and trophic adaptations;
- **Opportunists** - who feed from an unusual source in their diet, or take advantage of an abundant or unusual food source.

Gerking (1994) explains that an opportunist can also feed on some other food source when the present one is heavily exploited or suffers a seasonal decline. However, another term could be added, as I understand that the most appropriate word would be the term "Facultative", since opportunists can discard a common food source whenever another more advantageous one appears. The facultative feeding behavior of fish would be:

- **Facultative** - which feed on different types of food source, i.e. when there is a change in food items, due to high consumption by the aquatic community, and a sharp drop or scarcity of their preferred type of food in the environment. And they don't exhibit a behavior of taking advantage of some food (opportunism).

Since many ichthyans are generalists, the rest of the discussion focuses on specialists and opportunists, as well as facultatives. It should be noted that almost all fish are opportunistic in their feeding habits, except for the restricted specialists; and although these concentrate their energies on a restricted type of food and rarely expand their diet, it should be taken into account that most specialized diets may show some flexibility in their composition (GERKING, 1994; ABELHA; AGOSTINHO; GOULART, 2001).

In the context of feeding behavior, detritivorous fish are identified, for example, as:

- **Generalist Detritivore**

The fish of the Loricariidae family (order Siluriformes) are sometimes considered to be detritivores (many species), and there is some controversy about this; so much so that a review study in 2012 categorized them as generalist detritivores and herbivores (PEREIRA; RESENDE, 1998; GERMAN et al., 2010; LUJAN; WINEMILLER; ARMBRUSTER, 2012; OLIVEIRA; ISAAC, 2013). As herbivores, another study reports a subdivision into more specific categories for locarid species, such as algivores (algae), periphytivores (periphyton), granivores (seeds), and xylophages (wood); in addition to already being cited as iliophages (SALVADOR-JR; SALVADOR; SANTOS, 2009).

- **Detritivoro Specialist**

- Slime-eaters

The detritivorous-iliophagous fish are in the families Prochilodontidae, Curimatidae, and Citharinidae (for example, the species of the genus *Citharinus*), which belong to the order Characiformes; and in the family Cyprinidae, including the members of the genus *Labeo* in Africa and Asia, and the genus *Cirrhinus* in Asia (BOWEN; GU; HUANH, 2006; MOREIRA, 2007; PERESSIN; GONÇALVES; BRAGA, 2012).

Olaosebikan and Nzeh (2008) carried out a study on two African detritivores, *Labeo senegalensis* and *L. coubie*, on the anatomical and histological characterization of the digestive tracts.

- **Detrimental Optional**

The fish known as Tilapia have been shown to respond to changes in their environment through facultative feeding; several examples of studies commenting on their trophic plasticity have prompted speculation that at least some populations of Tilapia may have evolved to take advantage of a wide variety of food resources, in which they have displayed omnivorous, herbivorous or detritivorous habits (DOUPÉ; KNOTT, 2010).

Bowen, Gu and Huanh (2006) cited species of the genus *Tilapia* as Iliophagous.

Zaganini, Vidotto-Magnoni and Carvalho (2012) considered *Oreochromis niloticus* to be a detritivore species (detritus was a constant in the diet of almost all size classes) and *Tilapia rendalli* to be a herbivore, in the Barra Bonita reservoir (Tietê River, Sao Paulo, Brazil); these authors also mentioned that *Oreochromis niloticus* has also been considered a detritivore in reservoirs in Colombia (CALA; BERNAL, 1997), and an omnivore in Lake Paranoà (Distrito Federal, Brazil) (STARLING et al., 2002); and the species *T. rendalli* was defined as a herbivore in the Irai reservoir, in the state of Paranà (ABILHOA, 2005) and in the Zambezi River, in Africa (WINEMILLER; KELSO-WINEMILLER, 2003), but showed a detritivore diet in the Tibagi River, in Paranà (OLIVEIRA; BENNEMANN, 2005).

The species *Dorosoma cepedianum* (Family Clupeidae) consumes zooplankton and phytoplankton, as well as other life forms when they are available, and consumes benthic detritus when plankton is scarce; It is identified as an omnivore of suspension food, or detritivore of suspension food, in inhomogeneous distributions of tiny particles (< 1 mm), and also as having facultative suspension feeding (MUNDAHL; WISSING,

1988; HEIDMAN et al., 2012; HOLLEY et al., 2015).

- **<u>Opportunistic Detritivore</u>**

> Detritivore-Invertivore (invertebrates)

The species *Satanoperca pappaterra* (Cichlidae) is a typical bottom-dwelling fish, which, although studies show that it has been characterized as a detritivore-invertivore, feeding on a mixture of detritus and benthic organisms (invertebrates), has been shown to only include types of food in its diet as long as they are associated with the substrate, and it seems that they are selected according to their abundance in the environment (HAHN; CUNHA, 2005; SILVA; NATALI; HAHN, 2012).

Cassemiro et al. (2008) comment that research into the availability of food in the environment is essential to support the conclusion that *S. pappaterra* is an opportunistic species.

2.6 The Curimatid *Steindachnerina notonota*

The fish of the Curimatidae family (Toothless characiforms), in general, are popularly known as saguirùs, birus or manjubas, and comprise eight genera, which are widely distributed in the main river basins of the Neotropical region; They are distinguished from all other Characiformes mainly by the absence of teeth, a characteristic of adults, since teeth are only present in the juvenile stage (PERESSIN; GONÇALVES; BRAGA, 2012; QUEIROZ et al., 2013).

The species of this family have various body modifications associated with food acquisition and processing, for example, the loss of oral teeth, modifications to the mouth, the surface of the palate, the gill arches and the digestive tract; these adaptations are related to the species obtaining the food that is commonly present near the substrate at the bottom of aquatic environments (NETTO-FERREIRA; VARI, 2011; QUEIROZ et al, 2013).

The *Steindachnerina* genus is the second largest in terms of number of species among the Curimatidae genera, with 24 known species (Figure 8) (NETTO- FERREIRA;

VARI, 2011). There are already some genetic studies of curimatids, involving some species of the genus *Steindachnerina* (OLIVEIRA et al., 2002; BRASSESCO et al., 2004; SANTOS et al., 2006; DE ROSA et al., 2007; VENERE et al., 2008; MARTIN et al., 2009; SAMPAIO et al., 2011; SAMPAIO et al., 2016).

The species *Steindachnerina notonota* (Miranda-Ribeiro, 1937), popularly known in Brazil as saguirù, piabussù, piabaçu, sabarù, saburù, biruba and branquinha (Figure 9); It is an endemic species of the Brazilian semi-arid caatinga, i.e. found in rivers, lakes and reservoirs in the Northeast region, mainly in the states of Cearà, Rio Grande do Norte, Paraiba, the Maranhà-Piaui ecoregion and the central-northeast region of Brazil (SILVA; GURGEL; SANTANA, 2005; MONTENEGRO et al., 2011). This species of freshwater fish is found in important river basins in Rio Grande do Norte: the Apodi/Mossoró river basin, the Piranhas/Assu river basin, the Punaù river basin and the Pirangi river basin (NASCIMENTO et al., 2014).

The curimatid *S, notonota*, despite its small size and low commercial value, is commonly consumed by the local population and is of great ecological importance, as a link in the food chain, i.e. it represents an important food resource for carnivorous fish (including those that are economically relevant), as well as taking part in the recycling of organic matter, thus purifying bodies of water subjected to organic pollution (PEREIRA; RESENDE, 1998; AMORIM-TEIXEIRA; GURGEL, 2004; MONTENEGRO et al., 2011; PESSOA et al., 2013).

The literature records the work of Azevedo et al. (1938) on the general biology of the species and its internal meristic and anatomical characters, and the studies of Amorim-Teixeira (2002) and Amorim-Teixeira and Gurgel (2004), who reported on the feeding behavior and nutrient dynamics of this species in the Riacho da Cruz reservoir, in Rio Grande do Norte.

However, in relation to studies on this species, existing studies on the feeding of *S, notonota* classify their feeding habits based solely on the occurrence of the items present in their diets, without considering the morphological aspects of the digestive tract. The most recent study of this species was carried out by Silva, Gurgel and

Santana (2005), in which morphohistological analyses of the digestive tract of *S, notonota* were made. However, an analysis of the complete morphology of the digestive tract, including the various structures that make up the buccopharyngeal cavity and other parts of the tract, has not yet been described in the literature.

In view of the above-mentioned considerations and the scarcity of studies focused on animal morphophysiology, it is important to carry out an anatomical and histological characterization of the entire digestive tract of *Steindachnerina notonota*, as this will provide subsidies for future morphological and ichthyofauna studies, including its application as a promising species as an environmental bioindicator, as it also consumes pollutants present in the bottom mud of dulcicolous aquatic environments. Martinez, Guedes and Simonato (2008) and Meto et al. (2015) *have already* carried out studies with detritivorous fish, assessing the involvement of contamination in aquatic life, in relation to levels of pesticides (insecticides, fungicides, herbicides and acaricides) and toxicity to exposure to diesel oil, which caused functional damage to organs such as gills, liver, muscles and viscera.

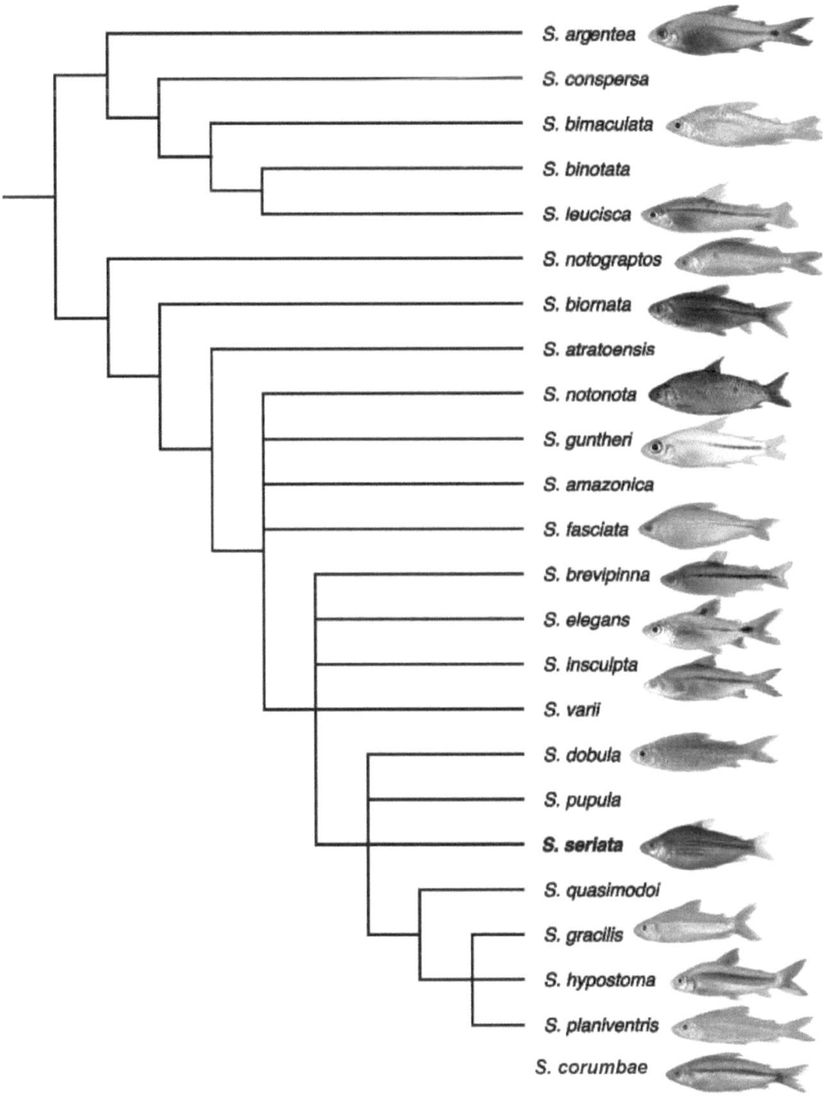

Figura 8 - Hypothetical cladogram of the phylogenetic relationships between *Steindachnerina* species (*S. corumbae* is not included). See Lucinda and Vari (2009) for a detailed explanation of these relationships.

Source: NETTO-FERREIRA; VARI, 2011; PROJETO SACI, 2016 (Available at: <http://www.projeto-saci.com/curimatidae/sem-subfamilia/steindachnerina-notograptos>. Accessed on: May 12, 2016).

Megaclass Osteichthyes

Superclass Actinopterygii

Class Actinopteri

Subclass Neopterygii

Infraclass Teleostei

Megacut Osteoglossocephalai

Clupeocephala supercut

Otomorpha cut

Subcategory Ostariophysi

Otophysa Section

Superorder Characiphysae

Order Characiformes

Suborder Characoidei

Family Curimatidae

Genus *Steindachnerina*

Notonota species

Figura 9 - Species under study *Steindachnerina notonota,* with the popular name Piabussù in RN, and its geographical distribution (yellow dot on the map): South America - Brazil.

See the new phylogenetic classification of bony fish (based mainly on molecular data - version 3), which is available on the DeepFin.org website: https://sites.google.com/site/guilleorti/home/classification.

Source: NETTO-FERREIRA; VARI, 2011; BETANCUR et al., 2013; FISHBASE, 2016.

3 OBJECTIVES

3.1 General Objective

This study aims to analyze the morphological aspects of the digestive tract of a species of neotropical teleost fish: *Steindachnerina notonota*, in relation to its detritivorous-iliophagous feeding habits in a dulcicolous environment; thus contributing to the expansion of information about the characterization of the macro and microscopic anatomy of its alimentary canal; and possible structural variations between the specimens studied.

3.2 Specific Objectives

- Anatomically describe the digestive tract of *S. notonota*;
- Characterize the gill tracks using ultrastructural scanning;
- Histologically analyze the cephalic intestine (only the gill tracks of the second gill arch and the epibranchial organ), the anterior intestine (esophagus and stomach), the middle intestine (cecal sac, pyloric cecums, proximal segment and distal segment), and the posterior intestine;

4 MATERIAL AND METHODS

4.1 Location of the Collection Area

The specimens used in this study came from the Cearà-Mirim River, in an area located between the coordinates 5°37'47" S Latitude and 35°37'9" W Longitude (Figure 10); in the municipality of Taipu, located in the mesoregion of Leste Potiguar, in the state of Rio Grande do Norte-RN, in the northeast of Brazil. This municipality is located in the semi-arid region and is 50 km from the capital Natal (IBGE, 2014; FEMURN, 2016).

The Cearà-Mirim River makes up one of the state's main river basins, covering an area of 2,635.7 km^2, corresponding to around 4.9% of the state's area, with its source in the municipality of Lages-RN, and its mouth in the Atlantic Ocean, in the municipality of Extremoz-RN, with a course of approximately 162 km (IDEMA, 2009; IGARN, 2014).

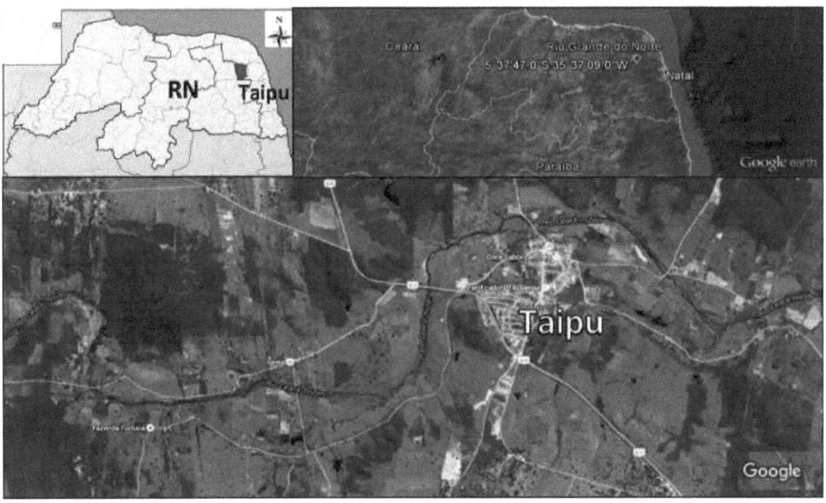

Figure 10 - GPS location of the collection area (red balloon): Municipality of Taipu, in the state of Rio Grande do Norte-RN (northeastern Brazil).

Source: GOOGLE EARTH /GOOGLE MAPS, 2016.

4.2 Specimen collection

In this study, 10 adult specimens of *Steindachnerina notonota* were captured, of both sexes, 5 females and 5 males, evidenced by the characterization of the gonads. The specimens were collected in August 2015, which is characterized as a very dry period (EMPARN, 2016), using a picaret and a 2 cm long net between opposite knots.

In the field, the live specimens were placed in a bucket with water from the river itself, and kept oxygenated by a portable battery-operated air pump compressor for aquariums (Jad D-200) (Figure 11), in order to survive and transport the fish to the Laboratory of Histological Techniques, in the Department of Morphology - CB/UFRN.

The species *S. notonota* has been deposited in the university's ichthyological collection under voucher number UFRN 902.

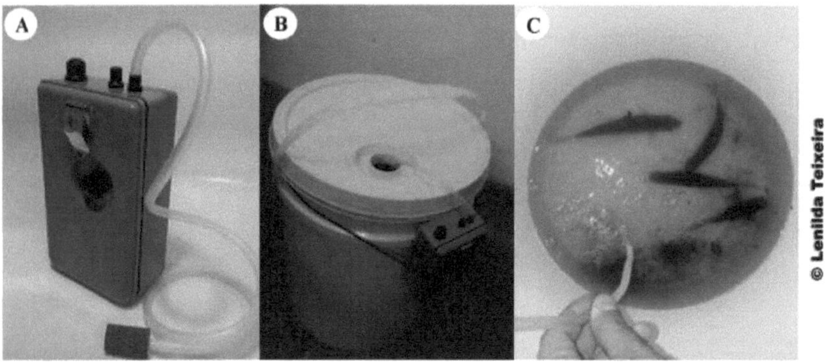

Figure 11 - Transportation and preservation of *Steindachnerina notonota* specimens. **A** - Aquarium battery-operated air pump compressor; **B** - Bucket with aquarium pump for transporting the collected specimens; **C** - Live specimens placed in the bucket with oxygen.

4.3 Bioethics Note

This study was approved by the Ethics Committee for the Use of Animals of the Federal University of Rio Grande do Norte, protocol No. 064/2015.

The euthanasia of the specimens collected was in accordance with the recommendations of the guidelines for the practice of euthanasia of CONCEA/MCTI (BRASIL, 2013); carried out in the Histological Techniques Laboratory of the Department of Morphology - CB/UFRN.

The fish were immersed in a buffered solution in a container containing the anesthetic benzocaine at a concentration of 100 mg/L of water for ten minutes or more, until equilibrium was lost and opercular movements ceased, followed by the physical method of sectioning the spinal cord.

4.4 Anatomical procedure

After euthanasia, the animals were measured and weighed, and biometry was carried out, recording individual data on total length (cm), standard length (cm) - distance from the tip of the snout to the end of the caudal peduncle (Figure 12), and total weight (g). The lengths were measured using a 6" - 150 mm x 1 mm metal caliper (Western), and the weight was checked using an electronic scale (AD330-Marte).

Then, a median ventral incision was made from the operculum to the anus, and up to the level of the lateral line, the gonads were removed, and with the exposure of the digestive tract structures, the specimens were preserved in 10% formaldehyde fixative.

All the structures were analyzed using a stereomicroscope - Leica M80 (Leica Biosystems) for macroscopic characterization of the structures, and photographs were taken with a Sony Cyber-shot dsc-w730 digital camera at the Cellular and Tissue Microscopy Laboratory of the Morphology Department - CB/UFRN.

Figure 12 - Measurement in centimeters of the total length (TL) and standard length (SL) of *S. notonota* specimens, using a caliper.

4.5 Histological Procedure for Light Microscopy

In the Histological Techniques Laboratory of the Morphology Department - CB/UFRN, after 24 hours of fixation, the biological material destined for histological processing was sectioned. The cephalic intestine (only the gill tracks of the second gill arch and the epibranchial organ), the anterior intestine (oesophagus and stomach), the middle intestine and the posterior intestine were sectioned and the fragments obtained were subjected to routine histotechniques (dehydration, diaphanization, inclusion and microtomy at 4 μm). These structures were stained using conventional histological protocols using Hematoxylin and Eosin (HE) (MAIA, 1979), and special, PAS - Periodic Acid Schiff, a mucoprotein-reactive method (BEHMER et al., 1976). After the slides were made, they were analyzed under a light microscope, where microphotographs of the slides were taken.

The microscopic images were obtained using a Zeiss scanner microscope: Axio Imager Z2, with a Metafer Vslide system for 8 slides (Metasystems, GmbH,

Germany), using 5x, 10x and 20x objectives; and a Nikon Eclipse 80i microscope, with Neurolucida software (MBF Bioscience, Williston, VT, USA), using 10x and 40x objectives; carried out at the Edmond and Lily Safra International Neuroscience Institute (IIN-ELS), in Macaiba-RN.

4.6 Procedure for Scanning Electron Microscopy

An ultrastructural study of the gill tracks of *Steindachnerina notonota* was carried out. After euthanasia, the gill tracks inserted into the buccopharyngeal cavity were removed and preserved in 10% paraformaldehyde. After fixation, the samples were dehydrated in increasing concentrations of ethyl alcohol (70%, 80%, 90% and 100%). After dehydration, the samples were sent to the Materials Structural Characterization Laboratory (Department of Materials Engineering/Centre of Technology/UFRN). The samples were placed on a metal support with the help of a double-sided carbon tape to fix them in place. The support was then placed in the sample holder in the SEM equipment (Model: Hitachi Tabletop Microscope TM-3000), for analysis and obtaining microphotographs of the material.

5 RESULTS

The specimens of *S. notonota* ranged in total length from 6.4 to 13 cm, standard length from 4.8 to 10.3 cm, and weight from 3.23 to 35.92g.

The animals were not found to have a marked sexual dimorphism in their body characteristics, such as coloration or fin shape, and sex was only confirmed from the presence of the female and male gonads.

The digestive tract of *Steindachnerina notonota* consists of the cephalic intestine (mouth and buccopharyngeal cavity), the anterior intestine (esophagus and stomach), the middle intestine (subdivided into the proximal intestinal segment and the distal intestinal segment, as well as the cecal pouch and the pyloric cecums), and the posterior intestine.

5.1 CEPHALIC INTESTINE

Mouth and Buccopharyngeal Cavity

- Lateral and anterior view

The mouth of *S. notonota*, located on the anterior portion of the head, is of the terminal type, i.e. the angle of the open mouth is forward (Figure 13). The oral rima, in anterior view, is characterized by the shape of an inverted "V", with the branches open and facing ventrally. It has a small amplitude of the oral cleft. The upper lip is well pigmented (several black dots scattered around a lighter area), facing slightly downwards, and is more developed (thicker) than the lower lip. In the center of the mandible there is a small rounded protuberance, which fits perfectly into the central groove of the maxilla. The mandible is foot-shaped (Figure 14). In a lateral view, the right and left ends of the maxilla (near the labial commissures) showed the presence of labial folds, which are loose, facing ventrolaterally and overlapping the ends of the mandible (Figure 14).

When the operculum is removed, the pharyngeal gill apparatus can be seen, located in the posterior portion of the head (Figure 13). The pharyngeal gill apparatus is made up of four pairs of gill arches, in which the gill tracks (located on the inner surface of the curvature of the arches) and the gill filaments (arranged on the outer surface of the curvature) are present (Figure 15).

Gill arches I, II, III and IV are made up of two branches: the upper, shorter, and the lower, longer; they decrease in size from the first to the last pair, as does the lower branch.

Under the stereomicroscope, with the arches separated and placed sideways, you can see the very small tracks, which are most noticeable in arches I, II and III.

Figure 13 - Lateral view of the mouth and pharyngeal gill apparatus of *S. notonota.*

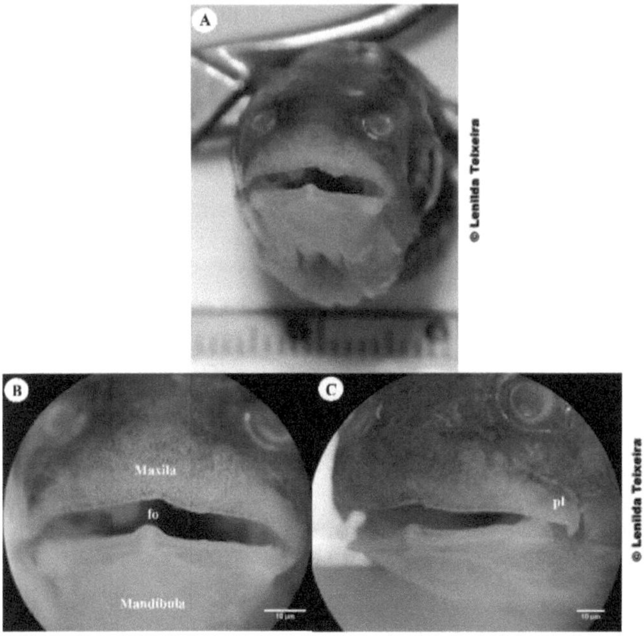

Figure 14 - Mouth of *S. notonota. A* - Frontal view; **B** - Frontal view at higher magnification, showing the oral cleft (fo); **C** - Lateral view showing the labial fold (pl).

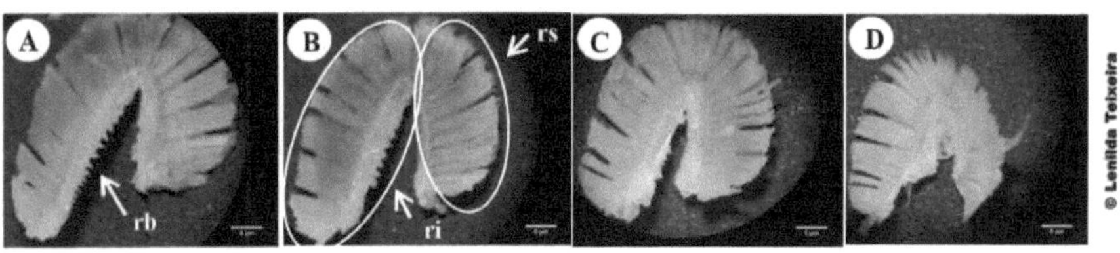

Figure 15 - The gill arches of *S. notonota.* **A** - arch I, with gill tracks (rb); **B** - arch II, showing the upper branch (rs) and the lower branch (ri); **C** - arch III; **D** - **arch** IV. Scale bar: 5 μm.

▪ Inside view

The buccopharyngeal cavity is characterized by the presence of the oral valve and a buccopharyngeal complex on the roof; with the structural absence of the tongue on the floor; it is completely devoid of teeth; it also has the pharyngeal branchial apparatus and the epibranchial organ (Figures 16 to 18).

The buccal region constitutes the anterior portion of the buccopharyngeal cavity and is located delimited cranially by the free edges of the lips and caudally by the first gill slit (both on the roof and on the floor).

The pharyngeal region is in turn made up of the posterior portion of the buccopharyngeal cavity, bounded cranially by the first branchial cleft (both roof and floor), and caudally to the inclusion of the epibranchial organ.

> Mouth region

On the buccal roof of the buccopharyngeal cavity (Figures 16 and 17), just after the upper lip, is the arch-shaped upper oral (or maxillary) valve, which is attached to the maxilla by means of the anterior or insertion edge; the posterior edge is free and projects into the buccal cavity, forming a barrier; it has a thin, smooth consistency.

The rest of the area is then occupied by the various soft tissue elaborations, which are modifications of the mucosal tissues, i.e. composed of the multiple lobulated fleshy bodies that make up the buccopharyngeal complex, which consists of its four main anatomical subunits:

(1) Anterior lobulated processes (ALP) - these are three longitudinally oriented wavy folds: a median fold and a pair of lateral folds, in the most anterior portion on the roof of the oral cavity;

(2) Posterior Iobulated Processes (PLP) - Located ventrally in relation to the orbits (right and left), they are fleshy protuberances and/or Iobulated bodies; along the entire surface of their edge facing the cavity, they have several relatively large papillae;

(3) Posteromedian flaps: anterior (APA) and posterior (APP) - these are fleshy flaps located in the region immediately anterior to the upper attachment area of the first gill arches; the APA has small fringe-like protrusions on its free edge, in which the protrusion in the center stands out because it is slightly larger;

(4) Lobulated processes of the first and second branchial arches (PLAB) - these are two bodies separated from each other by a small space, shaped like ovoid mounds, located in the proximal region of the upper attachment area of the first and second

branchial arches.

The floor of the mouth is characterized as a smooth surface.

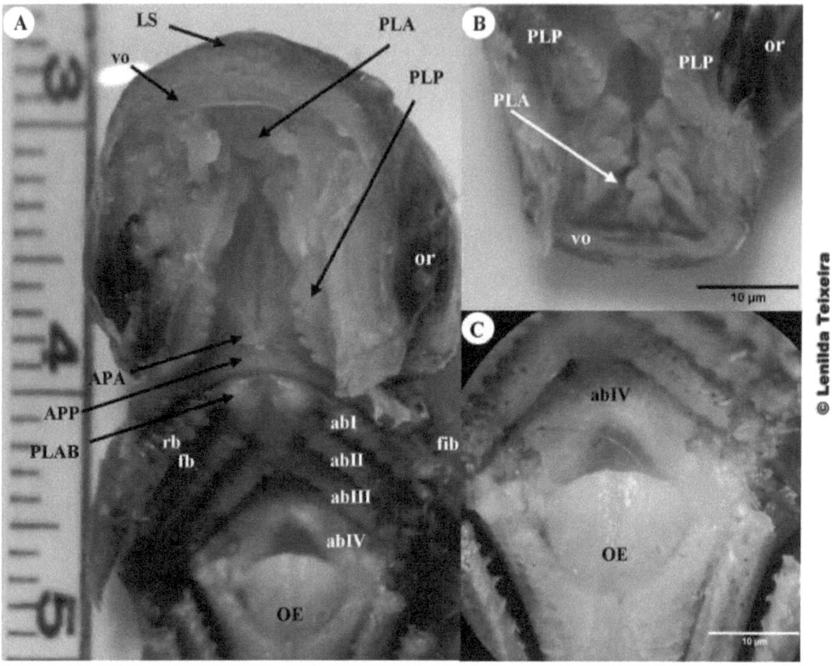

Figure 16 - Internal view of the roof of the buccopharyngeal cavity with indication of the structures. **A** - General view of the buccopharyngeal complex; **B** and **C** - Magnified view of some structures. **LS** - upper lip; **vo** - oral valve; **or** - orbit; **PLA** - anterior lobulated processes; **PLP** - posterior lobulated processes; **APA** - anterior posteromedian flap; **APP** - posterior posteromedian flap; **PLAB** - lobulated processes of the first and second branchial arches; **rb** - gill tracks; **fb** - gill slit; **fib** - gill filaments; **abI** - gill arch I; **abII** - gill arch II; **abIII - gill** arch III; **abIV - gill** arch IV; **OE** - epibranchial organ.

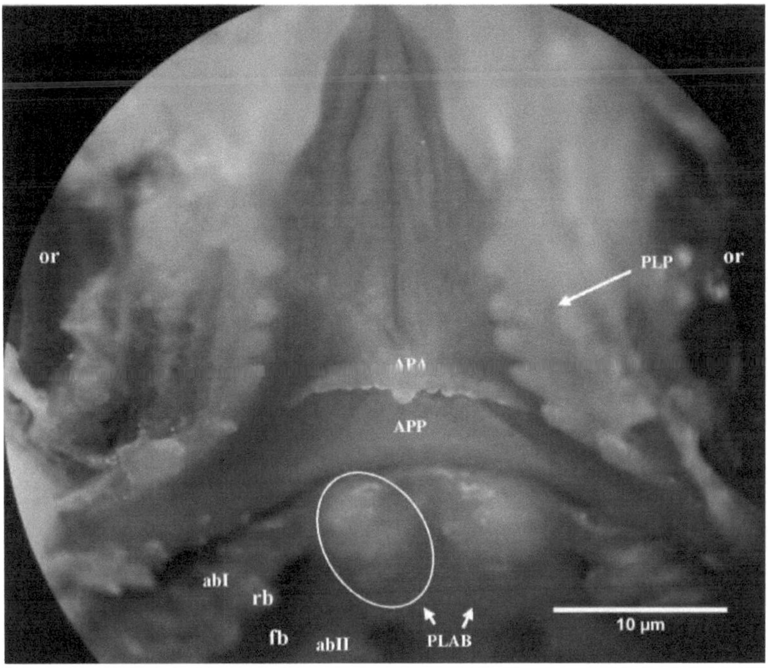

Figure 17 - Enlarged view of the buccopharyngeal complex with structures indicated. **or** - **orbit; APA** - anterior posteromedian flap; **APP** - posterior posteromedian flap; **PLAB** - lobulated processes of the first and second branchial arches; **rb** - branchial tracks; **fb** - branchial cleft; **abI** - branchial arch I; **abII** - branchial arch II.

> Pharyngeal region

The pharyngeal gill apparatus is located on both the roof and the floor (Figures 16 and 18). The upper branches of the gill arches are located on the pharyngeal roof and the lower branches on the pharyngeal floor; the epibranchial organ is located in a central space between the IV arches.

The upper branches of gill arches I, II, III and IV have two faces: the lateral (facing towards the operculum) and the medial (facing towards the inside of the cavity). Arches IV have lateral faces, but part of the medial faces are joined.

The arches are also formed by two edges: the anterior (inner surface of the curvature), where the gill traces are normally present, although no traces were found in the upper branch of arch IV; and the posterior (outer surface of the curvature), where the gill filaments are inserted. All the tracks are short and have an apparently gelatinous consistency.

Scanning electron microscopy showed greater detail, only the traces of arch II, corresponding to its upper branch (Figure 19).

The upper branch of arch II has two types of traces: those with a foliate shape, with a groove in the central region and the broad base is inserted in the anterior border, i.e. these tracks are all located in the anterior border, are evenly spaced, and occur in small numbers; on the medial face are the tracks with a pointed hook shape, arranged obliquely (with a curved direction), and are also evenly spaced and occur in small numbers.

Under light microscopy, these two tracks were lined with epithelium composed of sidewalk cells, with the presence of taste buds and mucous cells. In the pointed hook-shaped track, there are several taste buds arranged along its entire length; in the leaf-shaped track, fewer buds were observed (Figure 20).

On the pharyngeal floor of the buccopharyngeal cavity:

The lower branches of gill arches I, II, III and IV have two faces: lateral and medial. In arch IV, the lateral faces are free, but part of the medial faces are joined.

These arches are also made up of two edges: the anterior, where the tracks are, and the posterior, with the gill filaments.

Scanning electron microscopy showed the details of the tracks of the lower branches of arches I, II, III and IV. In the area of the pharyngeal floor, in the lower branches, we can see the gill tracks, of which there are a large number, always paired and together, and most of the time the lateral correspondents are joined with the medial ones at the anterior edges of the arches; at higher magnification, they have a conical shape with a recurved tip (Figure 19).

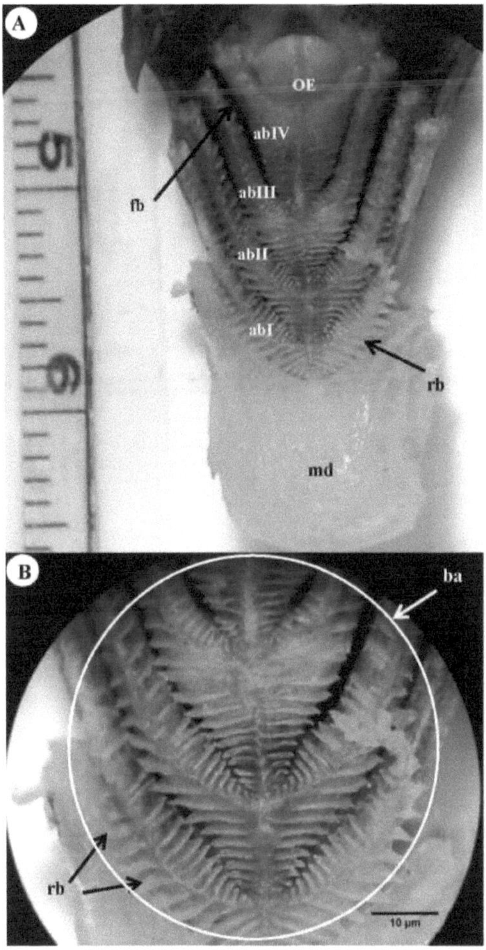

Figure 18 - Internal view of the floor of the buccopharyngeal cavity with indication of the structures. **A** - general view of the floor; **B** - enlarged view of the pharyngeal branchial apparatus. **md** - mandible; **rb** - branchial tracks; **fb** - branchial cleft; **abl** - branchial arch I; **abII** - branchial arch II; **abili** - branchial arch III; **abIV** - branchial arch IV; **OE** - epibranchial organ; **ba** - represents the plane of the anterior edges of each pair of branchial arches.

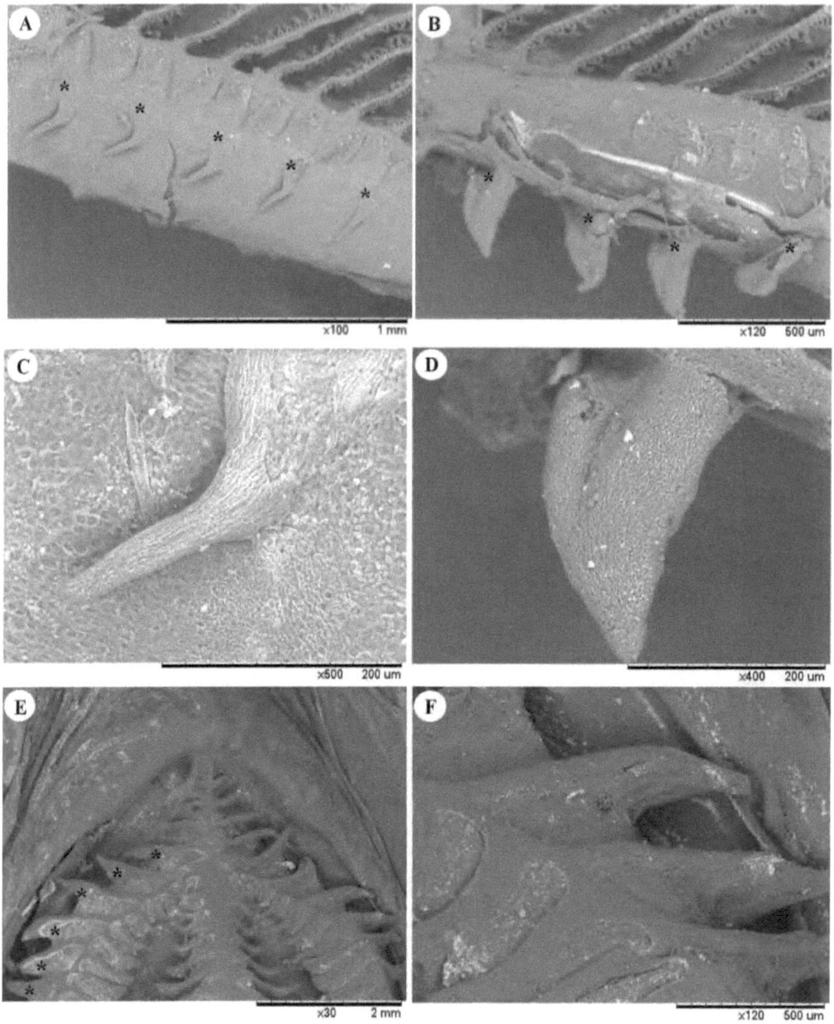

Figure 19 - Gill traces of *S. notonota. **A*** - Upper branch of gill arch II situated on the pharyngeal roof, showing the gill tracks (*) in the shape of a pointed hook on the medial side of the arch. **B** - Upper branch of gill arch II situated on the pharyngeal roof, the gill tracks (*) with a foliate shape, on the anterior edge of the arch. **C** - Gill track at higher magnification in fig. A. **D** - Gill track at higher magnification in fig. B. **E** - Area of the pharyngeal floor, in the lower branches, the gill tracks (*) inserted in the arches can be seen. **F** - Gill track in fig. E at higher magnification. SEM.

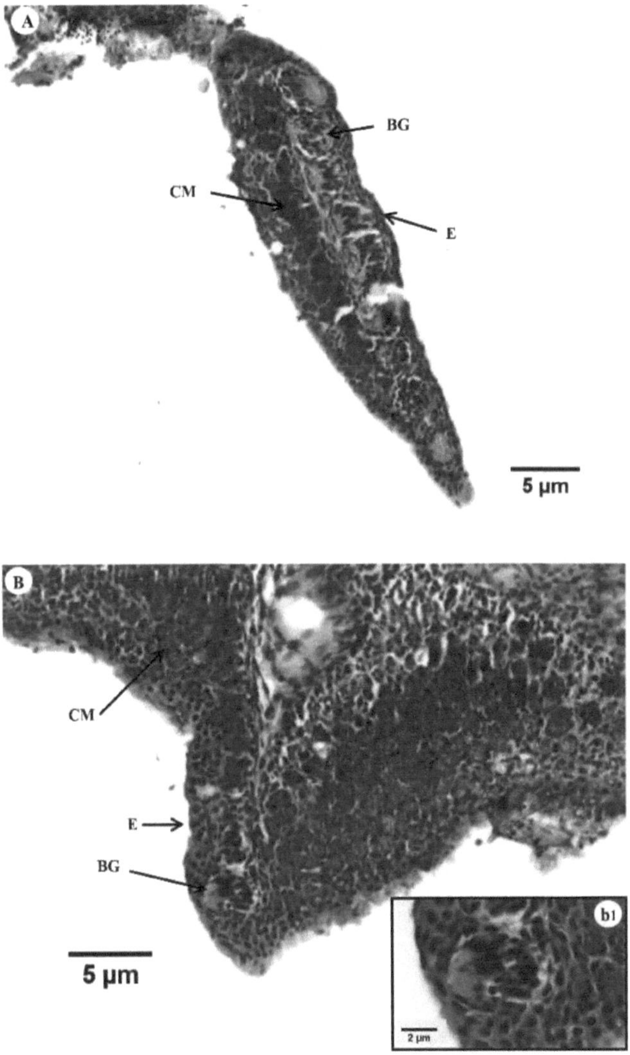

Figure 20 - Cross-section of the gill tracks of *S. notonota. **A*** - Pointed hook-shaped track; **B** - Leaf-shaped track. **CM** - Mucous cells; **E** - Epithelium; **BG** - Taste bud; **bl** - Enlargement of taste bud. HE.

The epibranchial organ of the pharyngeal region:

The epibranchial organ of *S notonota* is anatomically made up of two lobes joined by an isthmus (Figure 21).

In the posterior view, the right and left lobes are located between the last pairs of gill arches, lateral and ventral to them. They are located anterior to the esophagus. Each individual lobe is a diverticulum with a rounded tip. The lobes have a small space between them in the median part, which in most cases is quite far apart, and in some of them the separation is restricted to a single median groove.

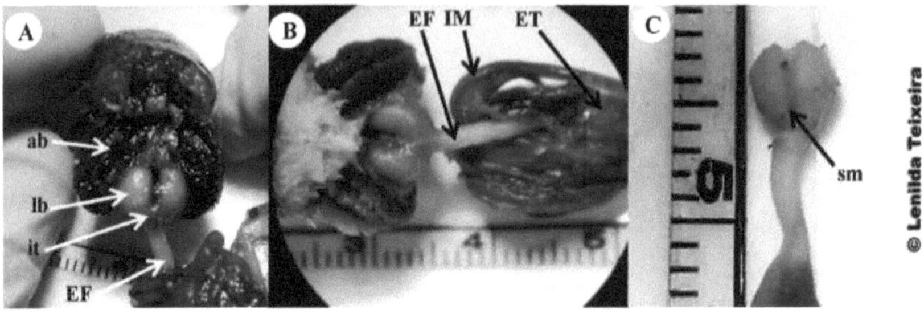

Figure 21 - The epibranchial organ of *S. notonota*. **A and B** - Posterior view of the organ connected to the last pairs of gill arches and the esophagus. **C**- Evidence of the median sulcus. ab - branchial arches; lb - lobe; it - isthmus; EF - esophagus; ET - stomach; IM - midgut; sm - median sulcus.

In cross-section, histologically (Figure 22 to 24), it is characterized by being an organ filled with connective tissue, with two types of channels: the entrance channel (present centrally throughout the entire length of the structure) and the tributary channels (two superior and two inferior). The entrance canal separates the upper and lower regions.

The epibranchial organ is joined to the arches by a hyaline cartilaginous center in the superolateral regions. The infero-lateral region has nerve bundles and blood vessels, nourishing the pharyngeal structure.

The last layer of tissue is a serous layer involving predominantly striated muscles in the ventral and dorsal regions. Its stroma also contains abundant adipose tissue. Internally, it has five areas with lumen (the canals), whose epithelia are distinct.

(I) Entrance canal: shaped like a concave mound with openings near the cartilaginous centers. They provide communication to the gill apparatus. The luminal mucosa of the canal has a pseudostratified mucosecretory epithelium with predominantly PAS-positive cells, taste buds and a lot of lymphocyte infiltration, supported by an obvious basal membrane. The lower portion of the canal mucosa has an epithelium superimposed on the lamina propria containing a thick adipose layer. The upper epithelium is supported by a thin lamina propria attached to the striated skeletal muscle.

(II) Tributary canals: these are two upper canals with a concave shape, in an incomplete ring, and two lower canals arranged obliquely with a rectilinear shape. They are all laterally oriented.

In the dorsal tributary canals, the upper part of the mucosa has papillae delimited by crypts lined with pseudostratified epithelium with mucous cells that react irregularly to PAS. The papillae have connective tissue, containing cells that are usually organized in a spiral, followed by the striated muscle layer; the lower part has a mucosa with pseudostratified epithelium with many PAS-positive mucous cells, which indicates the epithelium's great secretory action. Different from the epithelium in the upper region.

The same histological characteristics are observed in the ventral canals, but the positioning of the mucous membranes is reversed, with the papillae in the lower region.

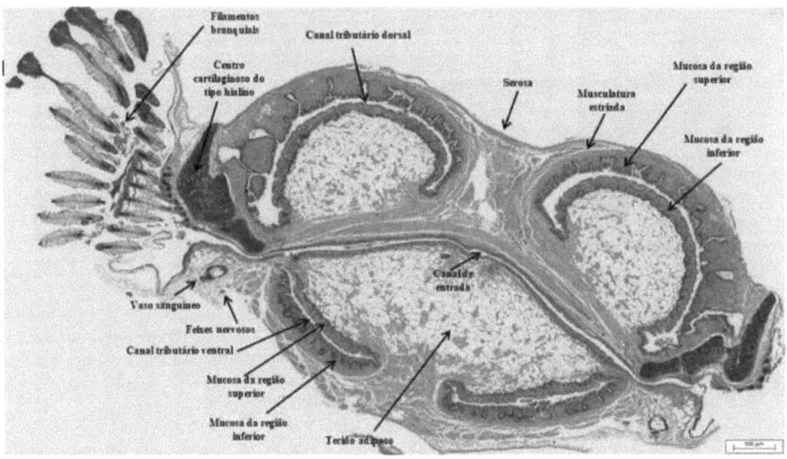

Figure 22 - Panoramic view of the epibranchial organ of *S. notonota*, in transverse section. HE.

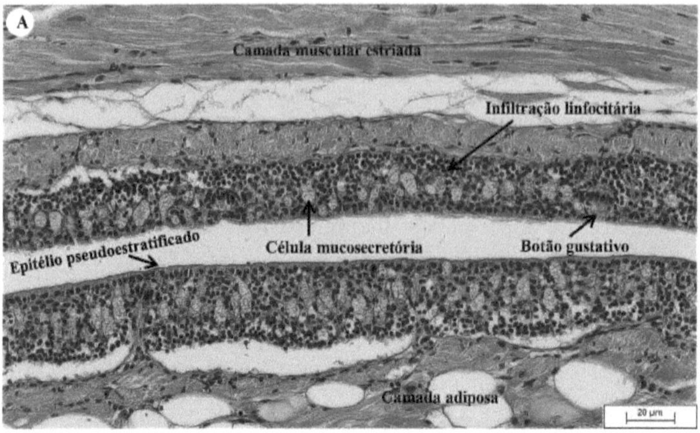

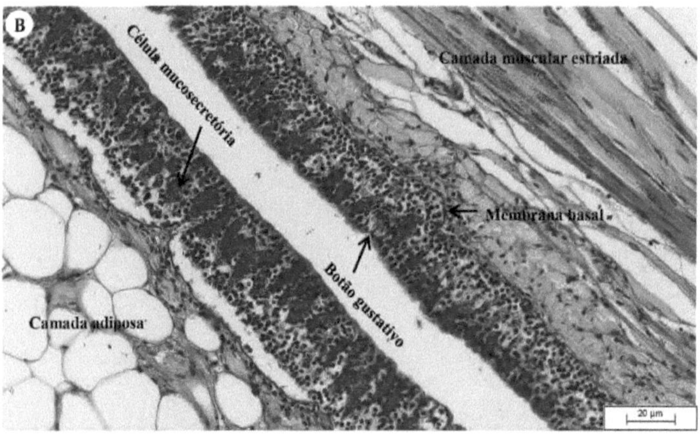

Figure 23 - Enlarged view of the entrance channel of the epibranchial organ with indication of the structures. **A** - HE staining; **B** - PAS staining.

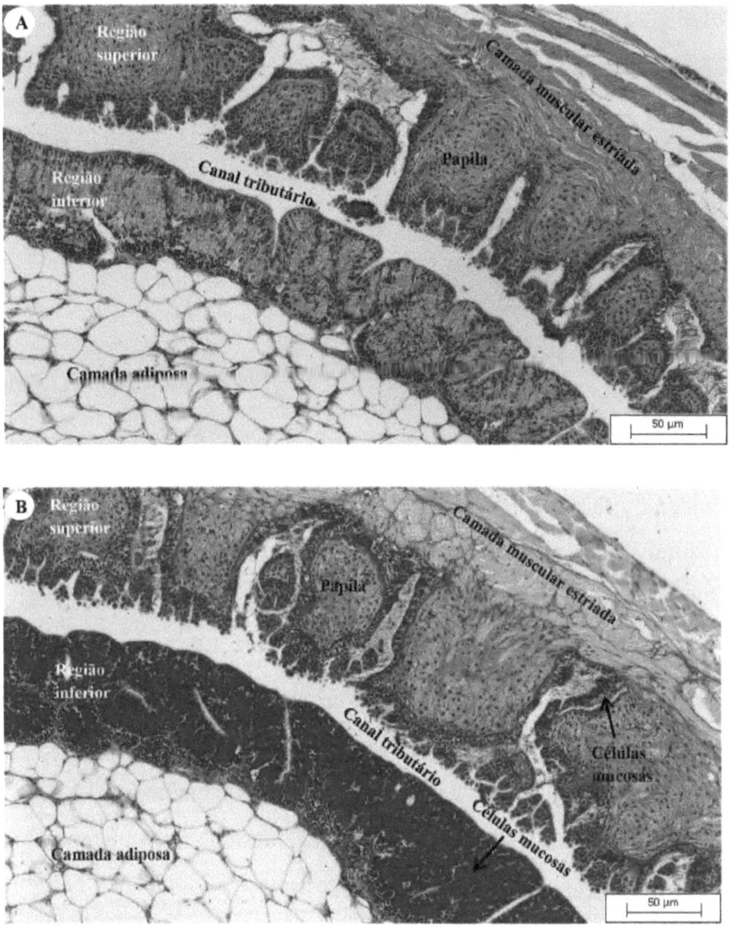

Figure 24 - Magnified view of the dorsal tributary canal showing the structures. **A** - HE staining. **B** - PAS staining.

5.2 PERITONEAL CAVITY

The peritoneal cavity of *S. notonota* is relatively small in size compared to its muscle mass (Figure 25).

The specimens analyzed were at an advanced stage of gonadal maturation. In the macroscopic characterization of the pairs of female and male gonads, they filled a large part of the peritoneal cavity, on both the right and left sides. In the females, the gonads were yellowish in color, very voluminous, with a large number of oocytes visible to the naked eye, and obvious blood vessels. In males, the gonads were also large and had a milky-white appearance.

Figure 25 - Adult specimen of *S. notonota*, showing the female gonad in the peritoneal cavity.

The organs of the digestive tract *in situ* are compressed by the gonads, the anterior intestine is surrounded by the middle intestine. This has its intestinal loops wrapped around itself in a spiral shape, and is situated more cranially in the peritoneal cavity and ventrally to the swim bladder, so the organs do not completely fill the cavity. The posterior intestine flows into the anus caudally, close to the anal fin (Figure 26).

In ventral view, the middle intestine with the anterior and middle parts being more voluminous than the posterior part; in which case, this anatomical conformation is greater in males than in females, as the pattern of coiling is greater and most of the time completely involves the anterior intestine (esophagus and stomach). Only by removing the organs from the peritoneal cavity and unwinding the intestinal loops is it possible to see the other components of the digestive tract (Figure 27).

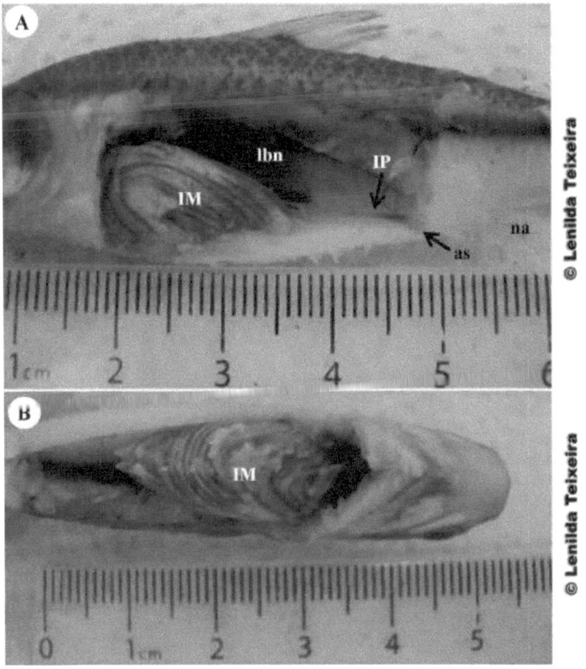

Figure 26 - Peritoneal cavity of *S. notonota*. **A** - Lateral view; **B** - Ventral view. **IM** - coiled midgut; **IP** - posterior gut; **lbn** - swim bladder site; **as** - anus; **na** - anal fin.

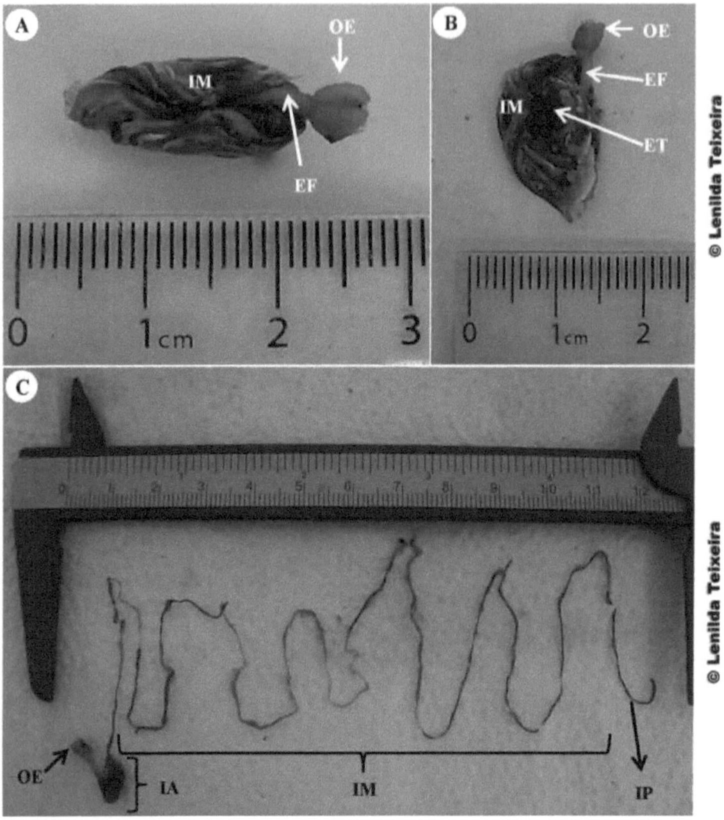

Figure 27 - Part of the digestive tract of *S. notonota* with structures indicated (A, B and C). **OE** - epibranchial organ; **EF** - esophagus; **ET** - stomach; **IA** - anterior intestine; **IM** - middle intestine; **IP** - posterior intestine.

5.3 ANTERIOR INTESTINE

Oesophagus

The esophagus is a tubular organ that connects the buccopharyngeal cavity to the stomach by connecting to the epibranchial organs, thus allowing food to pass through. It is a cylindrical structure, short and small in diameter.

The esophageal wall is thin, but elastic, and its internal surface has small longitudinal folds parallel to each other, continuing into the lumen of the organ. There is no sphincter between the esophagus and stomach, however, anatomically at this transition, there is a slight increase in width, and a change in color, with a yellowish color in the esophagus and brown in the stomach, and between the organs there is usually a green color, indicating food of plant origin (Figure 28).

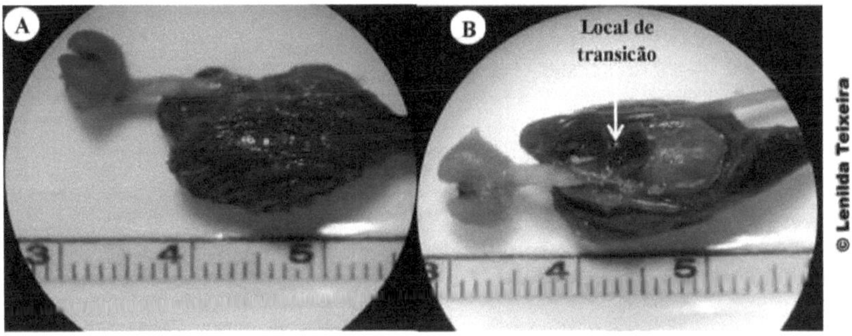

Figure 28 - Part of the digestive tract of *S. notonota*, showing the esophagus (A) and the transition site between the esophagus and the stomach (B).

Histologically, the mucosa that originates the esophageal folds is made up of a pseudostratified epithelium, and at no point is the muscularis of the mucosa evident (Figure 29). The epithelium lining the esophagus is made up of superficial cells, sometimes pavement-like, others mostly cuboidal, and numerous secretory cells. The secretory cells are located between the superficial cells, although in some sections they can be seen superimposed on them. The secretory cells are large and clear with a vertically flattened nucleus between the cell boundaries or in the basal portion of the cell. In the apical portion, pores open up between the superficial cells. These cells react regularly to PAS, and their secretions seem to be involved in moving food from the esophagus to the stomach because they are mucous.

The submucosa is aglandular and rich in fat cells.

The muscular layer of the esophagus is made up of two muscle sub-layers, an inner circular layer and an outer longitudinal layer, in both of which skeletal striated fibers predominate, but smooth muscle fibers will predominate in the posterior portion of the organ, close to the stomach.

Externally, the esophagus is surrounded by a serosa made up of loose connective tissue, with some nerve bundles and small vessels, delimited by the mesothelium.

Microscopically, the transition from the esophagus to the stomach is an abrupt replacement of the esophageal pseudostratified epithelium by the simple cylindrical epithelium of the stomach (Figure 30).

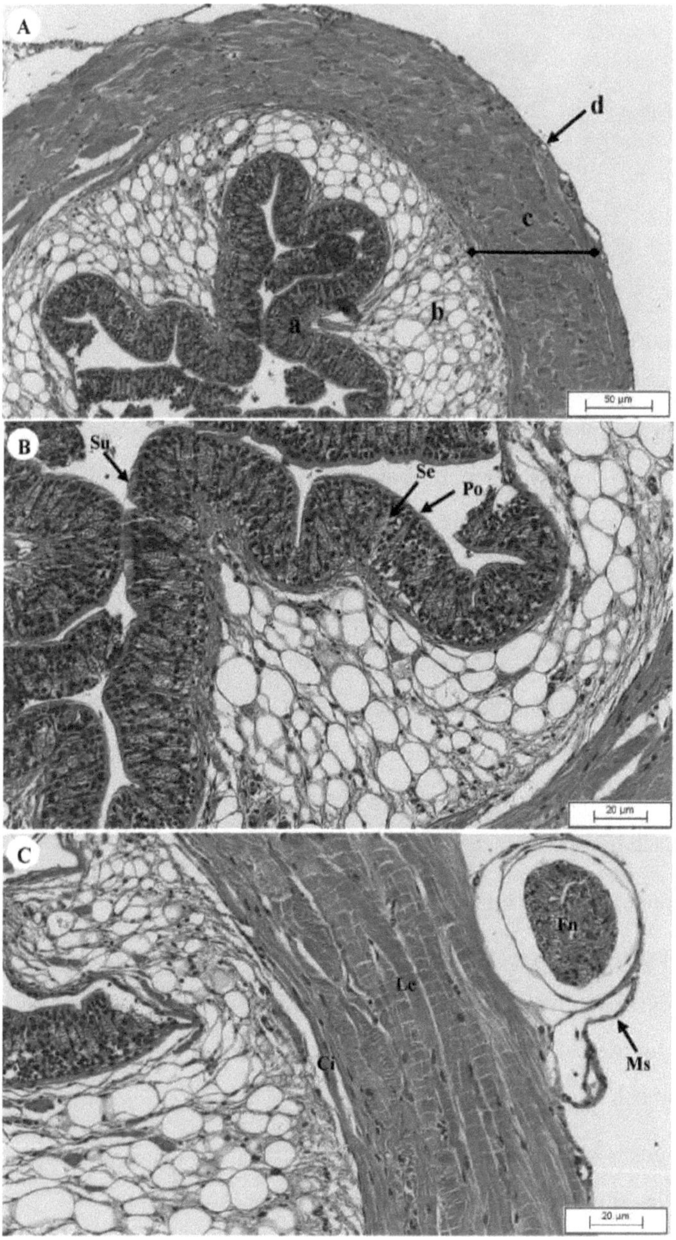

Figure 29 - Esophagus *of S. notonota. A* - Layers: **a)** Esophageal fold, **b)** Submucosa, **c)** Muscular, **d)** Serosa; **B** - Pseudostratified epithelium: **Su)** superficial cells, **Se)** secretory cells, **Po** - pore of the secretory cell. **C** - Muscular layer: **Ci)** inner circular, **Le)** outer longitudinal; **Fn)** nerve bundle; **Ms)** Mesothelium of the serosa. HE.

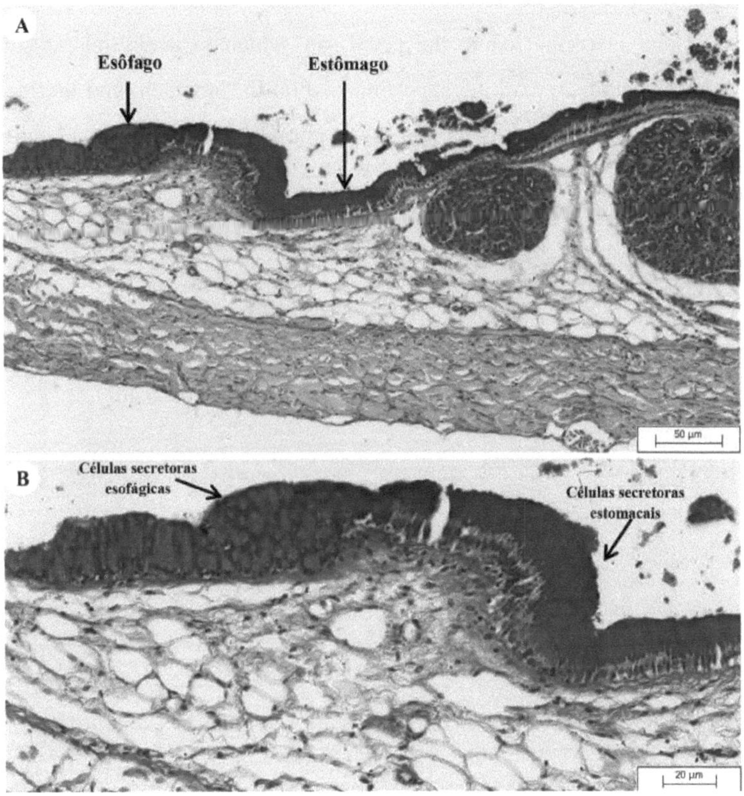

Figure 30 - Transition from esophageal to stomach epithelium of *S. notonota.* **A** - Esophageal pseudostratified epithelium and stomach simple cylindrical epithelium; **B** - Higher magnification showing the secretory cells of the esophagus and stomach in the cardiac region. PAS.

Stomach

The stomach is U-shaped and has three regions: the first, connected directly to the esophagus, has thin walls and a saccular appearance. It is called the cardia region and is responsible for initiating the digestion of food. The second, the fundic region, also saccular, is curved, with a blind bottom and more distended walls. It is also related to chemical digestion, but to a lesser extent than the cardia. The third region, connected to the fundic, has a periform shape (pear), very thick and rigid walls, and is the mechanical stomach (gizzard), called the pyloric region, which is involved in the grinding of food by breaking it down through mechanical action. In the cranial part of the pyloric region, there is a sharp constriction, which corresponds to the pyloric sphincter, whose region is connected to the cecal sac, which in its initial portion evaginates the pyloric cecums, remaining superimposed on the stomach; and its final portion undergoes a curvature (thick part) and then continues to the midgut (Figure 31).

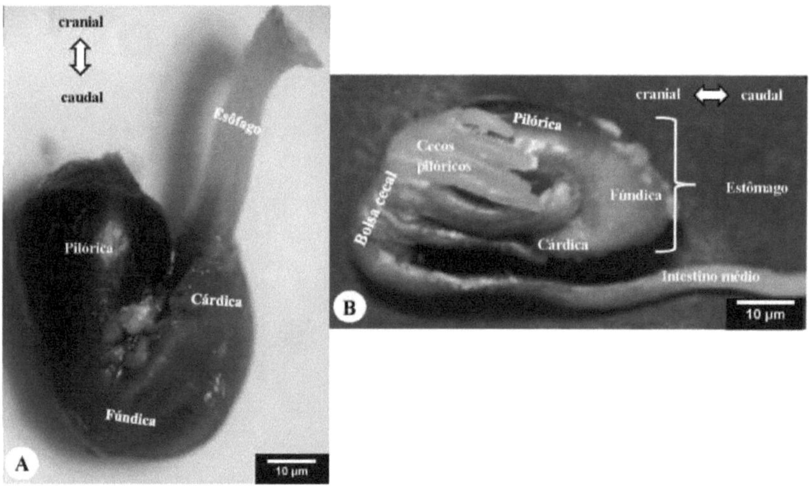

Figure 31 - Macroscopic images of the esophagus, the stomach regions and the beginning of the midgut, indicating the structures (A and B).

Histologically, the epithelial lining along the entire length of the stomach in *S notonota* is of the simple cylindrical mucosecretory type. The cylindrical secretory cells are predominant and homogeneously distributed throughout the lumen of the organ, matching each other in shape and size. They are PAS positive with clearly visible secretory granules located in the mid-apical portion. These cells have a spherical nucleus located in the basal portion of the cell. Gastric crypts or fossae are regularly arranged throughout the mucosa, differing in number, size and depth depending on the region considered. They are invaginations of the lining epithelium into the lamina propria which establish a connection with the secretory portion of the gastric tubular glands.

The gastric glands, when present, are always located in the lumina propria. The cells that make them up have not been classified due to the need for more specific staining.

The muscularis mucosae and submucosae of the stomach of *S. notonota* are not evident in any of the regions of this organ.

The muscular layer of the stomach, unlike the esophagus, has an inner layer with longitudinally oriented smooth muscle fibers and an outer layer with circularly oriented fibers.

External to the muscle layer is the serosa, made up of a thin, almost imperceptible layer of connective tissue, on which the mesothelium rests.

Each region of the stomach has a different histological organization, due to the development of physiological functions characteristic of each portion. Each region is therefore described separately below:

Cardia region - It has a highly developed mucosa, with numerous small gastric pits and long gastric tubular glands occupying the entire lamina propria. It has an internal longitudinal muscle layer and an external circular layer, covering it is the serous layer, which is poorly vascularized (Figure 32).

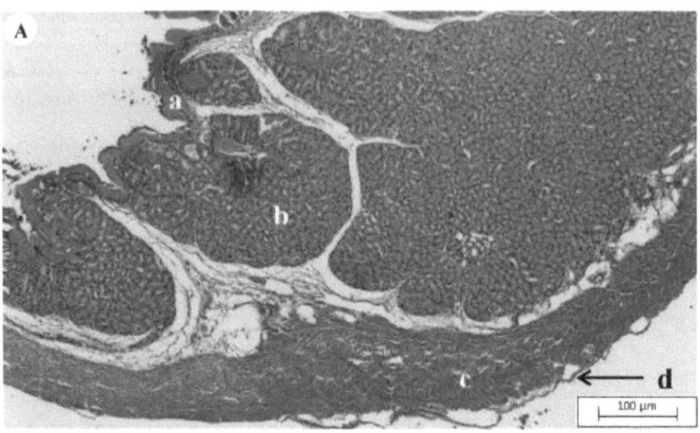

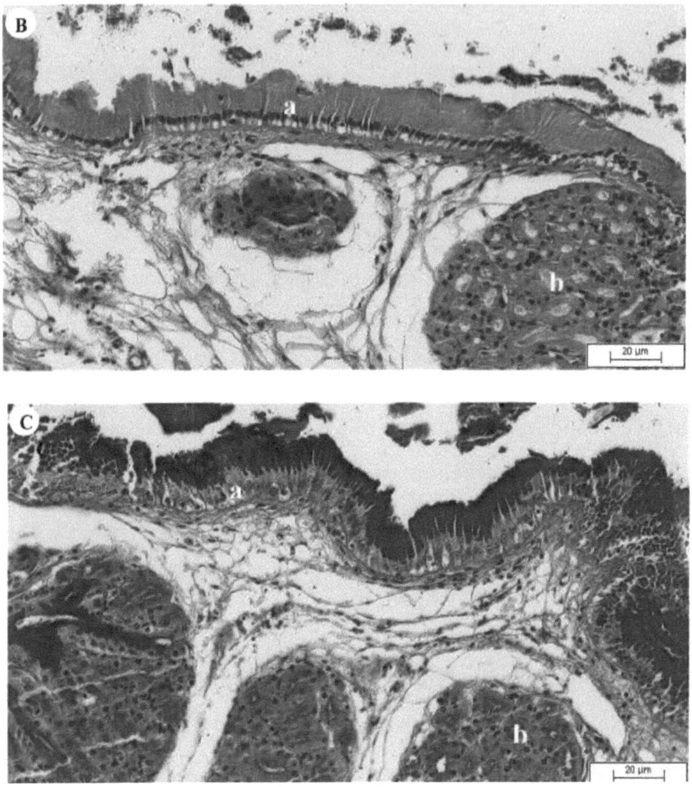

Figure 32 - Stomach cardia of *S notonota. **A and*** **B** - HE staining; **C** - PAS staining. **a)** Simple cylindrical mucosecretory epithelium, whose secretory cells are PAS positive; **b)** Tubular gland; **c)** Muscular layer; **d)** Serosa.

Fundic Region - The mucosecretory cells of its epithelium show high activity through PAS, to the point where a very thick layer of mucus is formed lining the lumen in this portion, which doesn't happen in the cardia region where there is little mucus. This layer of mucus becomes more pronounced in the pyloric stomach. In this region, the pits are deeper, while the gastric glands are shorter; both in smaller numbers than in the cardia region. The muscularis mucosae, submucosae and longitudinal muscularis were not visualized separately, but the very well-developed circular smooth muscle layer is very evident (Figure 33).

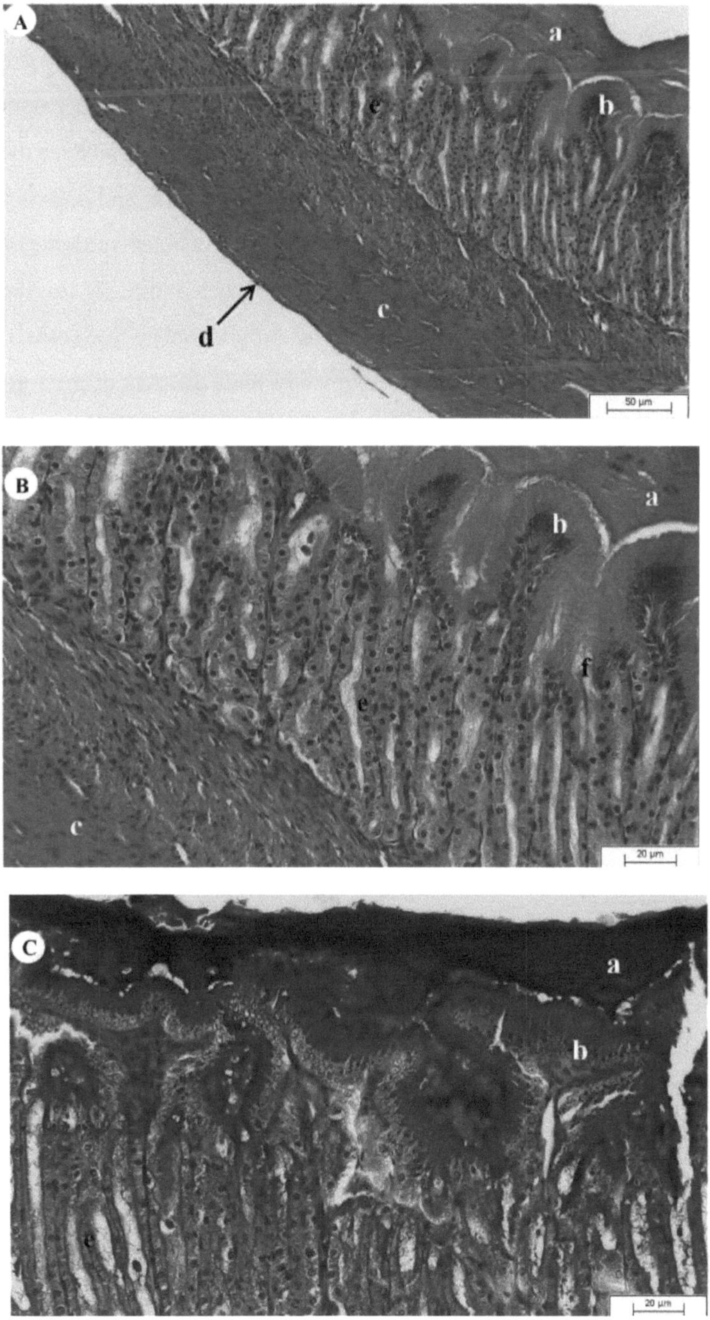

Figure 33 - Stomach fundic region. **A and B** - HE; **C** - PAS. **a)** Mucus layer; **b)** Cylindrical epithelium simple

mucosecretory (PAS-positive secretory cells); **c)** Muscle layer; **d)** Serosa; **e)** Gastric gland; **f)** Gastric fossa.

Pyloric region - It has an exceptionally more developed mucus layer covering the luminal surface than the previous regions. The mucosa of this portion, together with the circular muscle, forms four to six large longitudinal primary folds which reduce the lumen of the organ. The lining epithelium covers small secondary folds and does not give rise to crypts. There is no muscularis mucosae, but only the connective tissue of the lamina propria, which forms the secondary folds connected directly to the muscularis layer. There are no gastric glands in the lamina propria, only a few vessels. The hypertrophied muscular layer makes up approximately three quarters of the wall of the pyloric stomach and is formed predominantly of smooth muscle fibers arranged circularly with abundant vascularization. The longitudinal layer is not evident. A very thin serosa covers the organ (Figures 34 and 35).

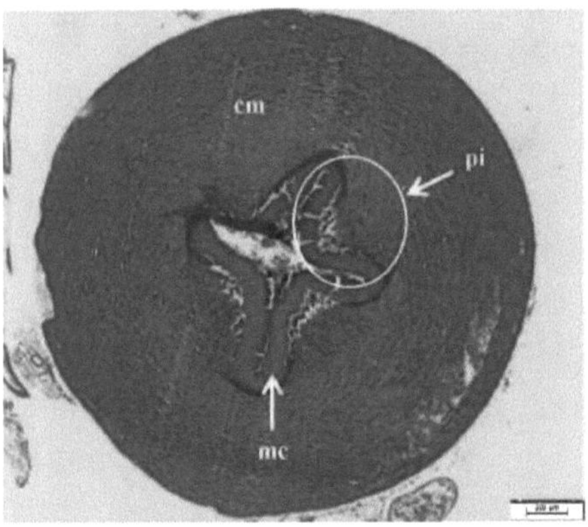

Figure 34 - Panoramic view of the pyloric region of the stomach of *S. notonota*. **mc** - mucus; **pi** - primary fold; **cm** - muscular layer. HE. Bar: 200 μm.

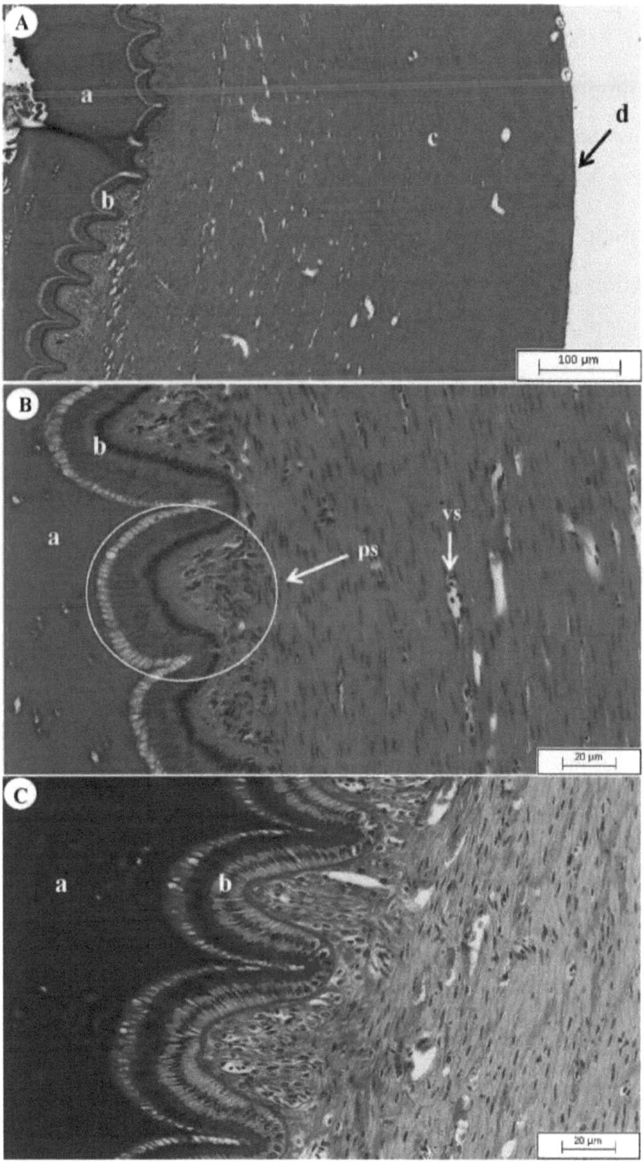

Figure 35 - Stomach pyloric region of *S notonota. **A and*** **B** - HE; **C** - PAS. **a)** Mucus; **b)** Simple cylindrical mucosecretory epithelium, whose secretory cells are PAS positive; **c)** Muscular layer; **d)** Serosa; **ps)** Secondary fold; **vs)** Blood vessel.

5.4 MIDGUT

The intestine is typical of iliophagous fish, very long and thin, intensely coiled, forming a spiral, with the circumvolutions of the intestinal loops joined by adipose tissue (Figure 36).

The midgut is continuous with the hindgut, forming a long cylindrical tube in which the morphological distinction is only evident microscopically. The caecal pouch and the pyloric caeca are part of the middle intestine.

Histological analysis of the midgut showed two types of regions, and it was subdivided into two segments: the proximal intestinal segment (the initial part continues into the cecal sac) and the distal intestinal segment (it is longer and continues into the posterior intestine).

Figure 36 shows a helical fold on the inside of the intestinal tube, which was observed in both the mid and hind intestines.

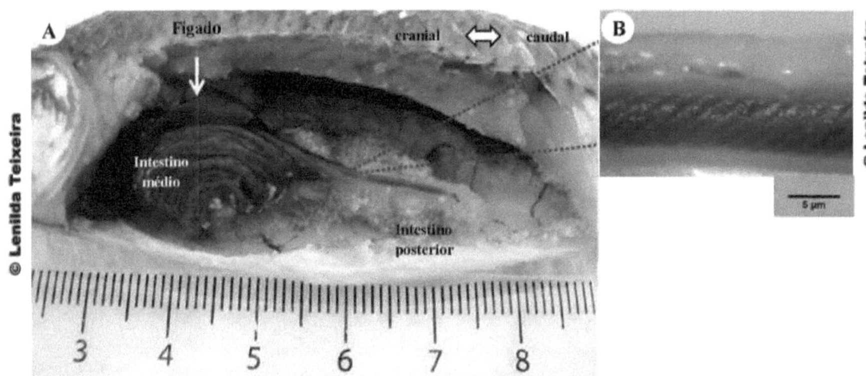

Figure 36 - Image of the midgut and hindgut (A) and the internal helical fold in the intestinal tube (B).

Cecal pouch

The cecal pouch has mucous folds and a simple cylindrical epithelium, with enterocytes and very few PAS-positive goblet cells. The mucous layer is located above the submucosa, after which follows the circular smooth muscle layer, and part of which is surrounded by the serosa. As the pouch is connected to the pyloric sphincter, it was possible to visualize the connection between these structures microscopically (Figure 37), but there may be a sharing of the outer muscle layer, as the adventitia was not visualized.

The pyloric sphincter has a simple cylindrical epithelium, with all the mucous cells in the apical part, which react positively to PAS, and no submucosal layer. The muscular layer is well developed.

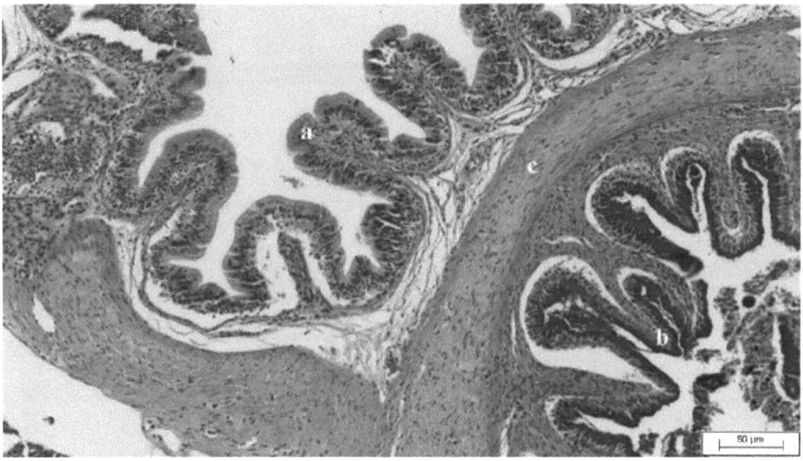

Figure 37 - Cecal pouch and pyloric sphincter *of S. notonota.* **a)** Epithelium of the cecal pouch; **b)** Epithelium of the pyloric sphincter; **c)** Muscular layer shared between the structures. PAS.

Pyloric cecums

The pyloric cecums, also known as intestinal cecums, are a set of small blind-bottomed tubular structures, varying in size from short to long, inserted cranially into the cecal sac. Their function is still much debated. For many authors, the caeca enlarge the internal surface of the intestine, increasing the absorption of nutrients.

Microscopically, an aglandular mucosa is observed, with long folds similar to villi, made up of a simple cylindrical epithelium and its own membrane of loose connective tissue, containing many lymphocytes and blood vessels (Figures 38 and 39). Specific PAS staining shows the epithelium with a few goblet cells interspersed in the enterocytes (Figure 39). The submucosal layer was not observed. The circular muscle layer is very thin and covered by serosa.

The pyloric caeca have the same type of epithelium as the intestine.

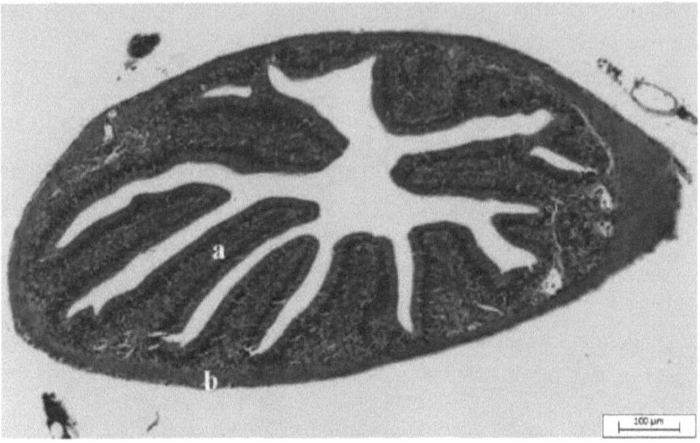

Figure 38 - General view of the pyloric caecum *of S. notonota*. **a)** Mucous layer; **b)** Muscular layer. HE.

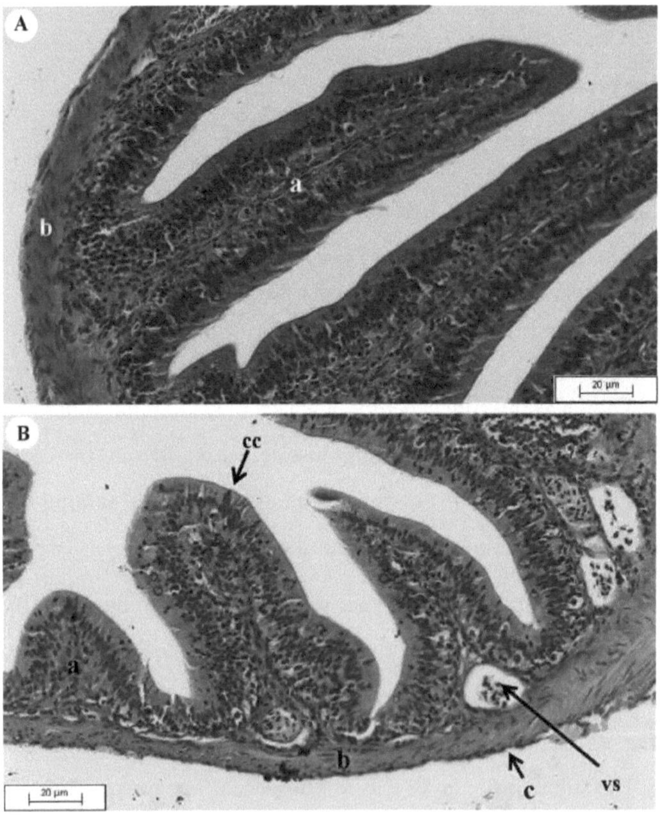

Figure 39 - Part of the pyloric caecum of *S. notonota*. ***A*** - HE staining; **B** - PAS staining. **a)** Mucous layer; **b)** Muscular layer; **c)** Serous layer; **cc) Goblet** cell; **vs)** Blood vessel.

Proximal Intestinal Segment

In the anterior segment of the midgut, the intestinal villi are lined with simple cylindrical epithelium, with enterocytes and rich in goblet cells that react positively to PAS. Compared to the distal segment of the midgut and the posterior intestine, the greater number of goblet cells in this proximal segment is noticeable (Figures 40 and 41).

The submucosa does not exist in any segment of the midgut. The circular smooth muscle layer, on the other hand, has bundles arranged longitudinally, with the presence of blood vessels, and is surrounded by the serous layer.

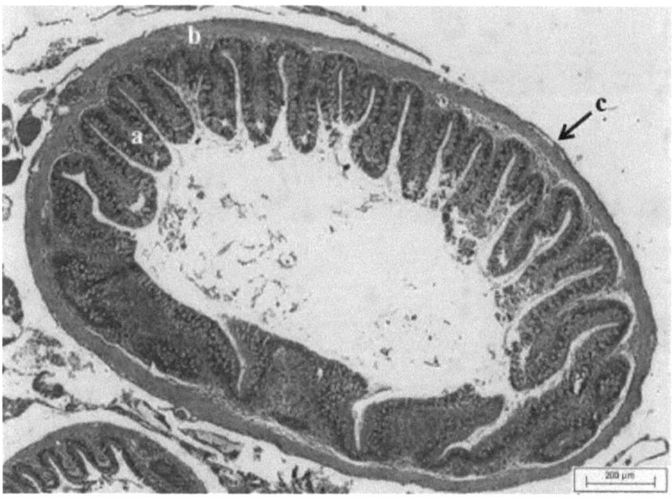

Figure 40 - General view of the proximal segment of the midgut *of S. notonota*. **a)** Mucous layer; **b)** Muscular layer; **c)** Serous layer. HE.

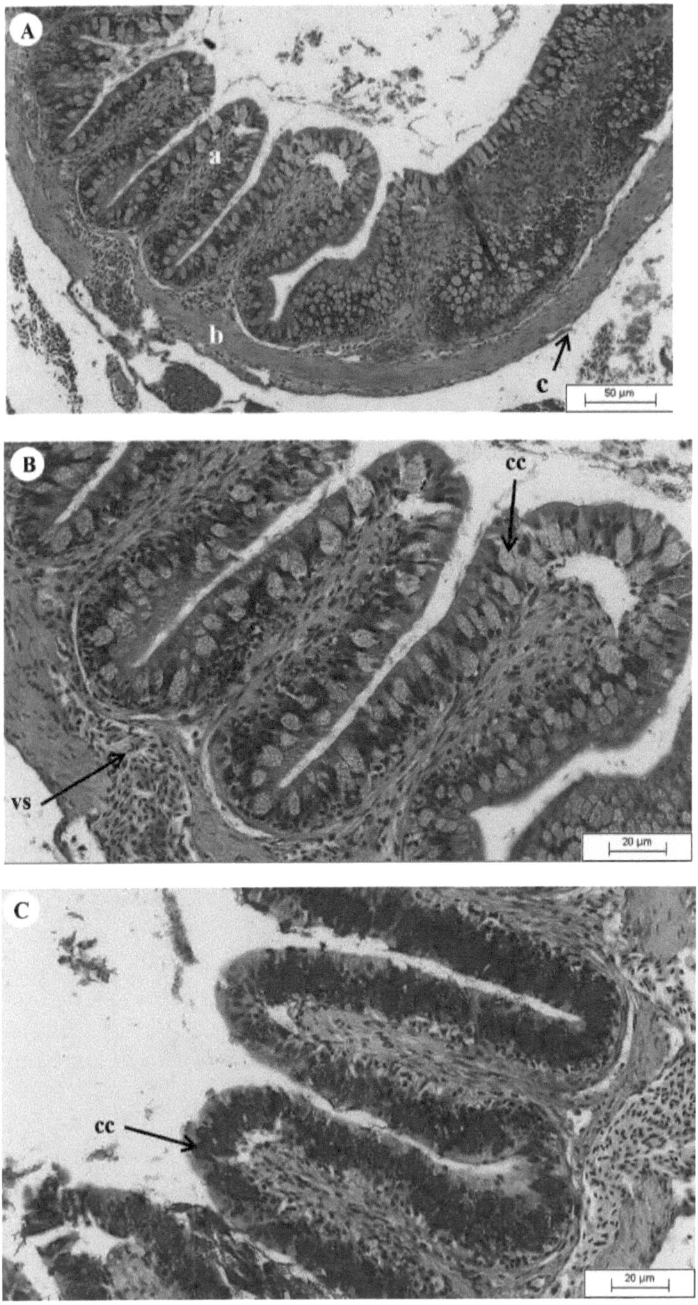

Figure 41 - Proximal segment of the midgut *of S. notonota. **A and*** **B** - HE staining; **C** - PAS staining. **a)** Mucous layer; **b)** Muscle layer; **c)** Serous layer; **cc) Goblet** cell; **vs)** Blood vessel.

Distal Intestinal Segment

In the midgut, especially in the distal segment, it is possible to visualize the helical fold macro- and microscopically.

It was possible to observe two areas, one without the helical fold and the other with the presence of this fold; both with mucosa containing villi lined by simple cylindrical epithelium, with few goblet cells compared to the previous segment, positive to PAS, and with a serous layer (Figures 42 to 44).

The area without the fold shows the intestinal villi and the circular and rather delicate longitudinal muscle layer, both of which are smooth muscles.

The area with the presence of the helical fold, in a longitudinal section, has the mucosal layer with projections of this fold, interconnected between the walls of the organ; and it was also possible to see the smooth muscle layer, which is thin and subdivided into internal circular and external longitudinal, which are clearly visible.

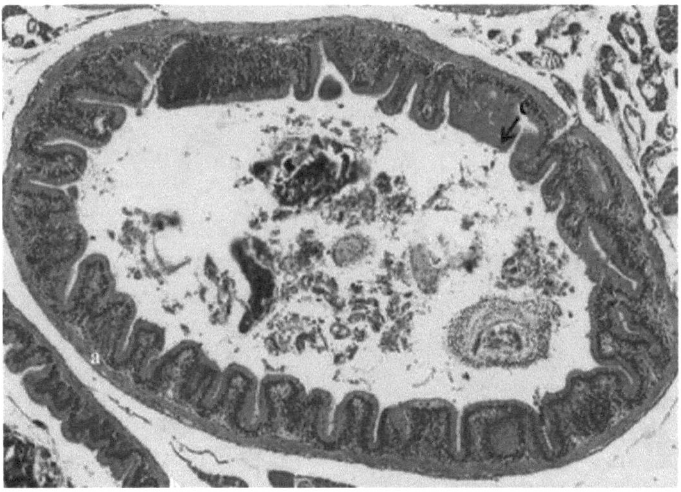

Figure 42 - General view of the distal segment (without helical fold) of the midgut *of S. notonota.* **a)** Mucous layer; **b)** Muscular layer; **c)** Serous layer. HE.

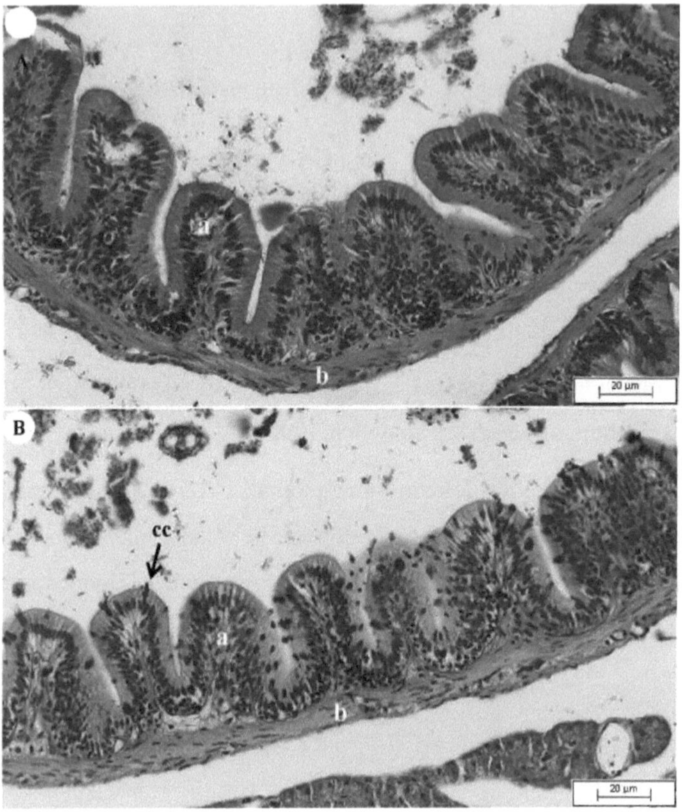

Figure 43 - Distal segment (without helical fold) of the midgut *of S. notonota.* **A** - HE staining; **B** - PAS staining. **a)** Mucous layer; **b)** Muscular layer; **cc)** Goblet cell.

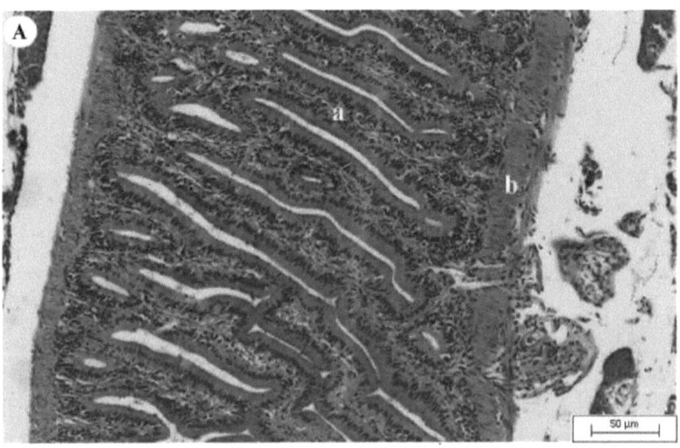

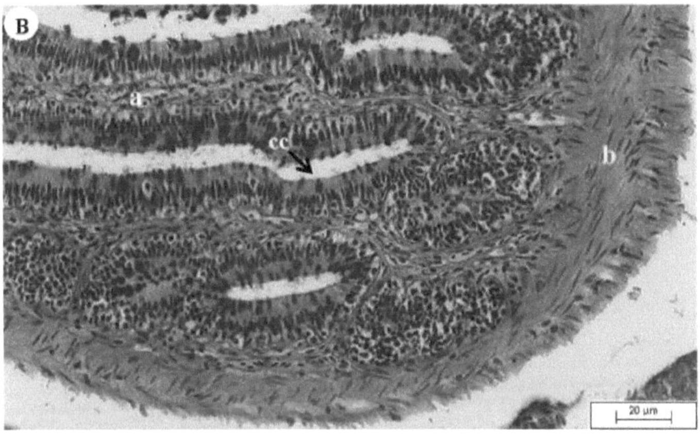

Figure 44 - Distal segment (with helical fold) of the midgut *of S. notonota*. **A** - HE staining; **B** - PAS staining. **a)** Mucous layer of the fold; **b)** Muscular layer; **cc)** Calyceal cell.

5.5 POSTERIOR INTESTINE

The posterior intestine Corresponds to the final part of the digestive tract and does not have an ileorectal valve.

Histological analysis showed no villi and intestinal folds with simple cylindrical epithelium and few goblet cells reacting positively to PAS; with aglandular submucosa, also with a very thin smooth muscle layer surrounded by serosa (Figure 45).

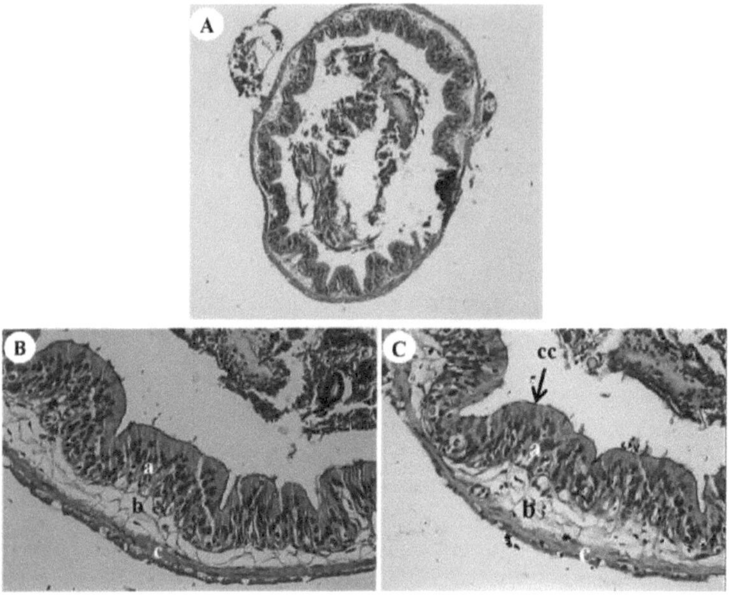

Figure 45 - Posterior intestine of *S. notonota*. **A** - General view. HE (10X); **B** - HE staining (40X); **C** - PAS staining (40X). **a)** Mucosal layer of the intestinal fold; **b)** Submucosal layer; **c)** Muscular layer; **cc) Goblet** cell.

6 DISCUSSION

DIGESTIVE TRACT

The anatomy of the digestive tract of fish follows the same basic plan as in other vertebrates with a degree of variation reflected in phylogeny and ontogeny, diet and environment (WILSON; CASTRO, 2011).

Fish are the most basal form of vertebrates, as they have some distinct characteristics, such as the absence of teeth and stomach in certain species, and undifferentiated small and large intestines; presenting various adaptations of their digestive system, according to the specialization required to ingest, digest and absorb the various types of food (LALL; TIBBETTS, 2009; CANAN et al., 2012; PESSOA et al., 2013).

The morphological characterization of the digestive tract was based on the following works: Menin (1988); Vari (1989); Vari (1991); Menin and Mimura (1991); Rodrigues and Menin (2005); Rodrigues and Menin (2006a); Rodrigues and Menin (2006b); Rodrigues, Navarro and Menin (2006); Makino (2010); Elsheikh (2013).

Buccopharyngeal cavity

The buccopharyngeal cavity is so called because it forms a single cavity with the buccal region and the pharyngeal region; and in the species *Steindachnerina notonota* it has various anatomical adaptations to the iliophagous feeding habit, i.e. ingesting the silt present at the bottom of dulcicolous aquatic environments.

Several authors have pointed out that the position of the mouth is related to feeding habits, to enable food to be obtained, whether on the surface, in the water column or at the bottom. According to Helfman et al. (2009), the vast majority of fish, regardless of their trophic habits, have terminal mouths, which means that the mouth opens at a forward angle, making it easier to take in food, which in this species promotes exploration of food on the bottom. The species *S, notonota* is the only one in the genus with a terminal mouth, as the other species have subterminal and inferior mouths (VARI, 1991).

The shape of the oral rima makes for a mouth that favors sucking in food, and the shape of the mandible allows the species to dig and take in sediment, which together with the presence of the labial folds and the oral valve helps to prevent the loss of soft sediment.

The terminal mouth of the detritivorous-iliophagous species *Prochilodus iineatus* is also adapted for sucking (MAKINO, 2010). In the literature, only Fugi (1993) reports the shape of the mouth of another species of the genus *Steindachnerina*, in which the mandible is also spade-shaped in specimens of *S, insculpta*. Also, in the study by Sazima and Caramaschi (1989), analyzing the feeding behavior of other curimatids, *Curimata spilura* and *C, nitens*, they described that these species swim close to the bottom and also use the mandible with this shape to dig and obtain sediment. And all adult members of the Curimatidae family lack oral teeth (GUISANDE et al, 2012).

In the species *S, notonota*, both the buccopharyngeal complex and the gill tracks promote the direction and containment of food in the buccopharyngeal cavity.

The gill trails have taste buds and mucous cells, and the epibranchial organ has taste buds in the lining of the entrance canal, as well as mucous cells in the tributary canals, so these structures are related to the chemoreception of food, and the mucus has the function of aggregating the fine particles of sediment, thus facilitating ingestion and forwarding to the esophagus. In this species, the

The characteristics of the epibranchial organ indicate that this structure participates in the digestive function.

The buccopharyngeal complex is described in a general way for all members of the genus *Steindachenrina*, in the work of Vari (1991), in which they mention the anatomical subunits that make it up, for systematic identification between species, and this complex is various soft tissue elaborations on the palate, found within the Curimatidae family, presenting some differentiations between species.

In teleosts, the pharynx has been characterized by the presence of gill arches, which present the gill tracks, which have been considered to play an important role in feeding, to aggregate particulate food, prevent the escape of food material, promote filtration and others (ELSHEIKH, 2013).

In the literature, there are still few histological and scanning electron microscopy reports of gill tracks. Elsheikh (2013) carried out a review on the ultrastructure of the surface of gill tracks, highlighting that studies have only been carried out on the following iliophagous species *Hypostomus commersonii* (Eiras-Stofella and Charvet-Almeida, 1997), *Prochilodus scorfa* (Eiras-Stofella and Charvet-Almeida, 1998), *Mugil curema, Mugil liza* and *Mugil platanus* (Eiras-Stofella et al., 2001), in which the shape and quantity varied between them and in relation to the species in this study.

Like *S. notonota*, the study by Lima et al. (2009), the species *S. brevipinna*, also presented arches with small gill tracks, lined by a stratified epithelium, with the presence of taste buds and mucous cells. And the authors Eiras-Stofella and Charvet-Almeida (1998), state that these taste buds have chemical receptors, related to the detection of food by the species.

The epibranchial organs will present a variety of morphological types in species that feed by filtration, known as microphages, because they consume fine particles of food; and there is a scarcity of studies describing their morphological characteristics, making it difficult to compare them with other species.

The exact function of the epibranchial organs remains enigmatic and is the subject of a centuries-old debate; many researchers have suggested respiratory, sensory and digestive functions; and in relation to the latter, in the species of *Prochilodus*, the function of the organs consisted of specialized filtration, thus relating them to the digestive function; in addition to the organs *of Dorosoma cepedianum*, *Dorosoma petenense* and *Coregonus lavaretus*, which were considered to be accessory structures to the digestive functions (MENIN, 1988; MENIN; MIMURA, 1991; HANSEN et al., 2014).

According to Menin and Mimura (1991), certain anatomical and histological aspects reported about these organs lead us to believe that in some species, these structures have a digestive function, due to the fact that they are more closely related to the digestive system than to the respiratory system; the innervation of these organs is carried out by the intestinal branch of the vagus nerve, and the blood supply comes from the dorsal aorta.

This study is the first macro- and microscopic description of the epibranchial organ of *S. notonota*, which showed many peculiarities, especially in relation to the types of epithelium present in its canals. It was only in the work of Menin and Mimura (1991) that this organ was reported in other truly detritivorous-iliophagous species, *Prochilodus marggravii* and *Prochilodus affinis*, in which, in comparison to the species *S. notonota*, the organs of the *epibranch* were more distinct. *notonota*, the organs are macroscopically located in the same position, that is, inserted in the last gill arches, and with the same shape in lobes joined by the isthmus, approached as a form of expanded primitive sacculus; and no histological analysis was carried out. However, Miller (1964) mentioned in his studies that the epibranchial organs had mucous cells, a cartilaginous center and striated muscles, as is present in *S'. notonota.*

It is thought that the epibranchial organs function as structures that collect and consolidate tiny particles of food. of food, i.e. they help by temporarily storing/accumulating the food in the form of a food bolus with mucus, which is then conducted to the esophagus by muscular action (HANSEN et al., 2014; HOLLEY et al., 2015). In the *S. notonota* species, this organ is possibly a structural adaptation of the digestive tract to the iliophagus habit.

Digestive tract

In the species in this study, histologically, the wall of the esophagus to the posterior intestine can present the following concentric layers, starting from the surface of the organ's lumen: mucosa, submucosa, muscle and serosa. The criterion used to differentiate the mucosal layer from the submucosal layer was based on the idea that they should be distinct tissue layers. According to Wilson and Castro (2011), the muscular layer of the esophagus is composed of skeletal muscle rather than smooth muscle. Most of the characteristics of the digestive tract corroborated the work of Silva, Gurgel and Santana (2005), since in this study a more comprehensive analysis was carried out.

Oesophagus

In his study, Menin (1988) found that the longitudinal organization of the esophageal folds is probably related to the distensibility of the walls of this organ, making it suitable for transporting food to the stomach. He also proposed that the pattern of the mucosa, whether more or less pleated, would lead to a greater or lesser capacity for swallowing unfractionated food. However, in *Prochilodus marggravii* and *P. affinis*, which feed on organisms present in the inert mass they swallow, these folds are less numerous and are probably more functional in directing food to the stomach than in esophageal distension. These observations corroborate the results found in this study for *Steindachnerina notonota*, taking into account the small number of longitudinal folds and the reduced esophageal lumen.

Chaves and Vazzoler (1984) found that the *oesophagus of Semaprochilodus insignis* comprises two microscopically distinct regions, the anterior lined with stratified sidewalk epithelium and the posterior with simple cylindrical epithelium. The *S notonota* species has only pseudostratified epithelium containing secretory cells with characteristics similar to those found in the esophageal epithelium *of Oreochromis niloticus* L. (MORRISON; WRIGHT JR., 1999), *Pimelodus maculatus* (GODINHO et al., 1970). Possibly, the pseudostratified epithelium has the function of protecting the esophageal mucosa against friction resulting from the passage of food through the esophagus.

The secretory cells of *S. notonota* regularly react to PAS. According to Al-Hussaini (1949) the secretion further aids in the agglutination of food particles and facilitates their transportation in an aboral direction.

Honda (1979) in his work with *Pseudocurimata gilberti* found that energy was transferred from somatic tissues to the development of the gonads. In this way, it can be suggested that *S. notonota* accumulates energy in the form of lipids present in the adipocytes of the esophageal lamina propria in order to use them later in the maturation of its gametes.

Stomach

Anatomically, the stomach of *S. notonota* is similar to those described by Fugi and Hahn (1991) and Chaves & Vazzoler (1984), with the fundic region being the curvature or joining loop, to which the authors refer respectively, due to its location and intermediate histological characteristics such as the thickness of the muscle and the number of gastric glands.

Menin (1988) reports a broad singularity for the fundic region, which can also be called the blind fundus, cecal region, joining bulb, cecum, blind sac, joining loop, gastric cecum, among others. Franco (1994) divided the stomach of the species he studied into three distinct regions: cardia, at the boundary with the esophagus; fundus, covering the middle portion; and pyloric, in contact with the intestine. Thus, the middle stomach region in the *S, notonota*, can be called the fundic region.

Franco (1994) found that even though *Hypostomus commersonii* is an iliophage, it has a stomach in the pyloric region, full of air, with a thin, translucent wall, lacking glandular structures, with a paved epithelium, and the presence of many vessels close to the organ's lumen. He concluded that for this species, the stomach has a digestive-respiratory function. In *S, notonota*, the opposite is true: the pyloric stomach has thick walls, which is probably related to the mechanical digestion of food.

The *S, notonota* species has a large number of gastric tubular glands in the cardia, a few in the middle and none in the pyloric region. It also has a muscular layer that gradually develops from the cardia to the pyloric region, reaching its maximum point in the pyloric region, where it takes on the shape of a gizzard.

Moraes et al. (1997), studying *Prochilodus lineatus* (Iliophaga), found a large number of gastric tubular glands in the cardiac region and no glands in the other regions, as well as an absence of mucosal muscle.

The pyloric region of the species in this study has no glands, which was also evident in the species studied by Franco (1994), Chaves and Vazzoler (1984), Menin (1988) and Morais et al. (1997).

The modification of the pyloric region into a mechanical stomach (hypertrophied circular layer) in the *S, notonota*, can be seen as an adaptation to the Iliophagus feeding habit. According to Nikolsky (1963) cited by Helfman et al., 2009, the organs for crushing and assimilating food are closely related to the variety of items consumed by the fish. Al-Hussaini (1949) considers the pyloric region, transformed into a gizzard, as in the Prochilodontidae, to be an adaptation to the need to break down and press ingested food.

Fugi & Hahn (1991) classified *Prochilodus scrofa* and *Curimata insculpita* as iliophagous based on the presence of a mechanical stomach (gizzard), while *Loricariichthys platymetopon* was characterized as detritivorous because it did not have one. This differentiation also took into account the occurrence and diameters of food items in the narrow pyloric lumen.

Iliophagous species feed on silt and have a large number of sand grains in their stomach contents, so they have structurally adapted stomachs that can withstand the friction caused by the grains. According to Amorim-Teixeira and Gurgel (2004), *S notonota* feeds mostly on sediment and algae, so this species has developed a mechanical stomach (gizzard) as a result of the type of food consumed in its population.

The pyloric sphincter marks the beginning of the intestine. Al-Hussaini (1946) for the iliophagous species *Milloides auriflamma* and Godinho *et al.* (1970) in *Pimelodus sp.* onivora, verified the presence of the pyloric sphincter. Thus, it can be suggested that the occurrence of the pyloric sphincter in the species has no correlation with feeding habits.

In this species, the cylindrical cells of the stomach epithelium are very reactive to PAS, with numerous apical granules being thrown onto the stomach wall, which form a very evident layer of mucus in all regions. According to Stoskopf (1993), this mucus layer is a thick layer of polysaccharides similar to keratin. It is therefore suggested that this layer of mucus protects the stomach epithelium against friction caused by the grains of sand present in the stomach contents of *S notonota.*

The muscular layer is responsible for gastric motility, and therefore for transporting and mixing food with stomach secretions. The circular muscular tunic in *S. notonota* makes up most of the pyloric wall, with fibers very close together, which creates a certain compactness in this layer. This feature, together with those of the epithelium, makes this region adapted to the iliophagous diet, and enables the species, which has no teeth, to grind food. Other species that are also iliophagous have this adaptation, such as *Semaprochilodus insignis* (CHAVES; VAZZOLER, 1984), *Prochilodus lineatus* (MORAIS et al., 1997), *Rhamdia branneri* (FRANCO, 1994), *Prochilodus scrofa* and *Curimata insculpta* (FUGI; HAHN, 1991), *Mulloides auriflamma* (AL-HUSSAINI, 1946).

Cecal pouch and pyloric cecums

The cecums, when present, can vary in number, shape and structure between species. The pyloric caeca are present in *S notonota* and vary in size. Drake et al. (1984), comments that in fish with short intestines there are more pyloric caeca. This relationship seems to exist in the species analyzed here.

The location of the pyloric caeca in *S. notonota*, inserted just after the pyloric sphincter in a serous pouch, was the same as that found in the studies by Chaves and Vazzoler (1984) for *Semaprochilodus insignis*, and Fugi and Hahn (1991) for *Prochilodus scrofa* and *Curimata insculpta*. The first authors reported that the caeca are surrounded by a connective tissue capsule and structurally seem to have no relation to nutrient absorption, as they lack absorptive structures, as seen in *Prochhilodus lineatus* by Angelescu and Gneri (1949). In the species in this study, due to the presence of absorptive and goblet cells, it is suggested that the caeca are involved in absorbing nutrients and lubricating the intestinal epithelium.

The caeca may increase the pH of the food bolus, making it alkaline (ZAVALA-CAMIN, 1996).

Structurally, the species in this study shows similarities between the caecums and the midgut, although there are fewer goblet cells in the former.

Middle and hind intestine

The intestine of *S, notonota* shows anatomical characteristics identical to those observed for the same species by Azevedo et al. (1938); for *P. scrofa and C. insculpta* studied by Fugi and Hahn (1991); and for *P, lineatus* analyzed by Morais et al. (1997). These authors refer to the long length and tangling of the intestine, relating it to the time needed for the difficult digestion of material such as sand, silt and cellulose.

The transport of food in the intestine of *S, notonota* is slower, due to the several intestinal loops and long length, as well as the presence of the helical fold, a structure only reported in this species so far, providing the advantage of food retention, i.e. slowing down the displacement of the intestinal contents, in order to exert greater nutritional absorption. This fold is similar to the spiral valve of elasmobranchs, which also greatly increases the internal surface area of the intestine.

Cartilaginous fish have an intestine with a spiral valve in different shapes between species, however, according to Romer and Parsons (1985), the primitive nature of the spiral structure is evidenced by its widespread presence in other groups of fish, thus, the spiral intestine occurs in chimaeras; and in fossils we can see its presence even in ancient placoderms; later, it was present in the first bony fish; living lungfish and all the lower actinopterygians retain the spiral valve, and even a few teleosts show traces of this structure.

Argyriou et al. (2016) report that the spiral valve is absent in teleosts, however, a vestigial structure was found in the species *S, notonota*, which corresponds to the helical fold, proving that this fish is an exception, as reported by Romer and Parsons (1985), which would be present in few teleosts.

Há there is a difference in the number of goblet cells between the proximal and distal segments of the middle intestine and the posterior intestine, where the proportion of goblet cells decreases towards the posterior intestine. The proximal segment of the midgut will have more goblet cells due to the fact that the food recently crushed by the pyloric stomach still contains many coarse particles, thus preventing abrasion during the passage of the food and protecting the lining epithelium.

Given the large number of lymphocytes found in the caeca and intestine of *S notonota*, it is possible that there is lymphocyte maturation in the intestine. Histophysiological studies are also needed to differentiate the areas of fat absorption, protein absorption and water and ion absorption.

7 CONCLUSIONS

Curimatids are adapted to a detritivorous iliophagous feeding habit, i.e. teeth are absent in adults, the species have long intestines and many gill tracks to provide greater efficiency in the use of fine particulate organic matter present in the soft sediment - silt (PERESSIN; GONÇALVES; BRAGA, 2012).

According to the findings of this study, it is clear that the morphology of the digestive tract is closely related to the detritivorous-iliophagous feeding habits of the species *Steindachnerina notonota*, whose structural adaptations are: a mouth that facilitates sucking in food; the buccopharyngeal complex - with various soft tissue elaborations on the palate, which provide direction and containment of the soft sediment; the presence of gill tracks and the epibranchial organ with its peculiarities, both with taste buds and mucous cells; a mechanical stomach with highly developed muscles in the pyloric region to grind the food; and a very long, coiled intestine, which has an exclusive helical fold on the inside to increase the absorption of low-calorie food that is difficult to digest.

REFERENCES

ABELHA, M. C. F.; AGOSTINHO, A. A.; GOULART, E. Trophic plasticity in freshwater fish. **Acta Scientiarum**, n. 2, p. 425-434, 2001.

ABILHOA, V. Ichthyofauna. In: ANDREOLI, C. V.; CARNEIRO, C. (Ed.). **Integrated management of eutrophicated water sources.** Curitiba: Gràfica Capital Ltda., 2005. p. 437-456.

AGOSTINHO, A. A.; PELICICE, F. M.; GOMES, L. C. Dams and the fish fauna of the Neotropical region: impacts and management related to diversity and fisheries. **Braz. J. Biol.**, v. 68, p. 1119-1132, 2008. Supplement 4.

AL-HUSSAINI, A. H. The anatomy and histology of the alimentary tract of the feeding coral, Mulloides auriflamma (FORSK) **J. Morph.**, v. 78, p. 121-154, 1946.

AL-HUSSAINI, A. H. On the functional morphology of the alimentary tract of some fish in relation to differences in their feeding habits. II - Cytology and Physiology. **Quart. J. Micr. Sec. London**, v. 90, p. 323-354, 1949.

ALVIM, M. C. C.; PERET, A. C. Food resources sustaining the fish fauna in a section of the upper Sao Francisco river in Très Marias, MG, Brazil. **Braz. J. Biol.**, v. 64, n. 2, p. 195-202, 2004.

AMORIM-TEIXEIRA, J. L.; GURGEL, H. C. B. Dynamics of nutrition and natural feeding of *Steindachnerina notonota* (Miranda-Ribeiro, 1937) (Pisces, Curimatidae), Riacho da Cruz Reservoir, Rio Grande do Norte, Brazil. **Rev. Bras. Zoo**, v. 6, n. 1, p. 9-28, 2004.

AMORIM-TEIXEIRA. J. L. **Seasonal and lunar variation in the feeding behavior of *Steindachnerina notonota* (Miranda-Ribeiro, 1937) (Pisces, Curimatidae) in the Riacho da Cruz reservoir, Rio Grande do Norte, Brazil.** 58p. 2002. Dissertation (Master's Degree) - Federal University of Rio Grande do Norte, Natal/RN, 2002.

ANDERSON, J. L.; CARTEN, J. D.; FARBER, S. A. Zebrafish Lipid Metabolism: From Mediating Early Patterning to the Metabolism of Dietary Fat and Cholesterol. **Methods Cell Biol.**, v. 101, p. 111-141, 2011.

ANDRIAN, I. F.; PERETTI, D.; BOSCO, P. C. D. Main morphological arrangements related to the feeding of herbivorous and carnivorous fish. **Arq. Apadec.**, v. 9, n. 2, p. 20-26, 2005.

ANGELESCU, V.; GNERI, F. S. Adaptaciones del aparato digestivo al régimen alimenticio en algunos peces del rio Uruguay y del rio de la Plata. **Rev. Inst. Invest. Mus. Argent. Cienc. Nat.**, v. 1, n. 6, p. 161-272, 1949.

ARGYRIOU, T.; CLAUSS, M.; MAXWELL, E. E.; FURRER, H.; SÂNCHEZ-VILLAGRA, M. R. Exceptional preservation reveals gastrointestinal anatomy and evolution in early actinopterygian fishes. **Scientific Reports**, v. 6, n. 18758, p. 1-9, 2016.

AZEVEDO, P.; DIAS, M. V.; VIEIRA, B. B. Biology of the Saguirù (Characidae, Curimatidae). **Mem. Inst. O. Cruz**, v. 33, n. 4, p. 481-553, 1938.

BECKER, A. G., GONÇALVES, J. F.; GARCIA, L. O.; BEHR, E. R.; GRAÇA, D. L.; KURTZ FILHO, M.; MARTINS, T.; BALDISSEROTTO, B. Morphometric parameters comparisons of the digestive tract of four teleosts with different feeding habits. **Ciência Rural**, Santa Maria, v. 40, p. 862-866, 2010.

BEHMER, A. O.; TOLOSA, E. M. C; NETO, A G. F. **Manual of techniques for normal and pathological histology**. Sao Paulo. Editora da Universidade de Sao Paulo, 1976, p. 98-101.

BEMVENUTI, M. A.; FISCHER, L. G. Peixes: Morphology and Adaptations. **Cadernos de Ecologia Aquàtica**, v. 5, n. 2, p. 31-54, aug.-dez., 2010.

BERTIN, L. Appareil digestif. In: GRASSE, P. P. **Traité de zoologie**, Paris, Masson, v. 13, p. 1249-1301, 1958.

BERTMAR, G.; KAPOOR, B. G.; MILLER, R. V. Epibranchial organs in lower teleostean fishes - an example of structural adaptation. **Int. Rev. Gen. Exp. Zool.**, v. 4, p. 1-48, 1969.

BETANCUR-R, R.; BROUGHTON, R. E.; WILEY, E. O.; CARPENTER, K.; LÓPEZ, J. A.; LI, C.; HOLCROFT, N. I.; ARCILA, D.; SANCIANGCO, M.;

CURETON, J. C.; ZHANG, F.; BUSER, T.; CAMPBELL, M. A.; BALLESTEROS, J. A.; ROA-VARON, A.; WILLIS, S.; BORDEN, W. C.; ROWLEY, T.; RENEAU, P. C.; HOUGH, D. J.; LU, G.; GRANDE, T.; ARRATIA, G.; ORTi, G. The Tree of Life and a New Classification of Bony Fishes. **PLoS Curr.**, v. 18, n. 5, 2013.

BLANCK, D. V.; TAKAKO, A. K. Current state of fish genetic improvement in Brazil. In: Marcos Tavares-Dias; Wagner dos Santos Mariano (Org.) **Aquaculture in Brazil: new perspectives.** Vol. 1. Sao Carlos: Pedro & Joao Editores, 2015. chap. 6, p. 119-138.

BLAZQUEZ, F. J. H. **Histology of the Digestive System of Teleost Fishes.** Lecture presented at the II Week on Fish Histology. Jabuticabal, Sao Paulo: FUNEP, 1995. p. 31-38.

BOWEN, S. H. Detritivory in neotropical fish communities. **Environmental Biology of Fishes**, Corvallis, v. 9, p. 137-144, 1983.

BOWEN, S. H.; GU, B.; HUANG, Z. Diet and Digestion in Chinese Mud Carp *Cirrhinus molitorella* Compared with Other Ilyophagous Fishes. **Transactions of the American Fisheries Society**, v. 135, p. 1383-1388, 2006.

BRAGA, R. R.; BORNATOWSKI, H.; VITULE, J. R. S. Feeding ecology of fishes: an overview of worldwide publications. **Rev Fish Biol Fisheries**, v. 22, p. 915-929, 2012.

BRAZIL. Ministry of Science, Technology and Innovation. **CONCEA euthanasia practice guidelines.** Brasilia: CONCEA, 2013. 54 p.

BRASSESCO, M.S.; PASTORI, M.C.; RONCATI, H.A.; FENOCCHIO, A S. Comparative cytogenetic studies of Curimatidae (Pisces, Characiformes) from the middle Paranà River (Argentina). **Genetics and Molecular Research**, v. 3, n. 2, p. 293-301, 2004.

BUCKUP, P. A.; MENEZES, N. A.; GHAZZI, M. S. (Ed.). **Catàlogo das espécies de freshwater fish of Brazil.** Rio de Janeiro: National Museum, 2007. 195 p.

CALA, P.; BERNAL, G. Ecologia y adaptaciones de la tilapia nilótica (*Oreochromis niloticus*) en ambientes naturales - Caso embalse de Betania y cienaga de chilloa, sistema del rio Magdalena, Colombia. **DAHLIA Revista Associaçâo Colombina de Ictiologia**, v. 2, n. 2, p. 3-29, 1997.

CALLAN, W. T.; SANDERSON, S. L. Feeding mechanisms in carp: crossflow filtration, palatal protrusions and flow reversals. **The Journal of Experimental Biology**, v. 206, p. 883-892, 2003.

CANAN, B.; NASCIMENTO, W. S.; SILVA, N. B; CHELLAPPA, S. Morphohistology of the Digestive Tract of the Damsel Fish *Stegastes fuscus* (Osteichthyes: Pomacentridae). **The Scientific World Journal**, v. 2012, p. 1-9, 2012.

CARRASSÓN, M.; GRAU, A.; DOPAZO, L.R.; CRESPO, S. A histological, histochemical and ultrastructural study of the digestive tract of *Dentex dentex* (Pisces, Sparidae). **Histol. Histopathol.**, v. 21, p. 579-593, 2006.

CASSEMIRO, F. S.; RANGEL, T. F. L. V. B.; PELICICE, F. M.; HAHN, N. S. Allometric and ontogenetic patterns related to feeding of a neotropical fish, *Satanoperca pappaterra* (Perciformes, Cichlidae). **Ecology of Freshwater Fish**, v. 17, n. 1, p. 155-164, 2008.

CASTRO, E. F. **Histological, histochemical and histochemical studies of endocrine cells in the stomach and midgut of freshwater fish (Teleostei) with different feeding habits.** 103 f. 2002. Dissertation (Master's Degree in Veterinary Medicine) - Federal University of Viçosa, Viçosa/MG, 2002.

CECAT - National Center for Research and Conservation of Cerrado and Caatinga Biodiversity. **Biodiversity - Caatinga.** Available at: <http://www.icmbio.gov.br/cecat/conservacao-dabiodiversidade/biodiversidade.html>. Accessed on: 24 Feb. 2016.

CHAVES, P. T. C. & VAZZOLER, C. Biological aspects of anatomical fish. III. Microscopic anatomy of the *esophagus*, stomach and pyloric cecum *of Semaprochilodus ensignis* (Characiformes: Prochilodontidae). **Acta Amazônica**, v. 14, p. 343-353, 1984.

CHAKRABORTY, C.; HSU, C. H.; WEN, Z. H.; LIN, C. S.; AGORAMOORTHY, G. Zebrafish: A Complete Animal Model for In Vivo Drug Discovery and Development. **Current Drug Metabolism**, v. 10, p. 116-124, 2009.

CHELLAPPA, S.; BUENO, R. M. X.; CHELLAPPA, T.; CHELLAPPA, N. T.; VAL, V. M. F. A. Reproductive seasonality of the fish fauna and limnoecology of semi-arid Brazilian reservoirs. **Limnologica**, v. 39, p. 325-329, 2009.

CHELLAPPA, S.; CÂMARA, M. R.; CHELLAPPA, N. T.; BEVERIDGE, M. C. M.; HUNTINGFORD, F. A. Reproductive ecology of a neotropical cichlid fish, *Cichla monoculus* (Osteichthyes: Cichlidae). **Braz. J. Biol.**, v. 63, n. 1, p. 17-26, 2003.

CHELLAPPA, S.; CÂMARA, M. R.; CHELLAPPA, N. T.; BEVERIDGE, M. C. M.; HUNTINGFORD, F. A. Reproductive ecology of a Neotropical cichlid fish Cichla monoculus (Osteichthyes, Cichlidae). **Braz. J. Biol.**, v. 63, n. 1, p. 17-26, 2003.

CHELLAPPA, S.; NASCIMENTO, W. S.; CHELLAPPA, T.; CHELLAPPA, N. T. Impacts of anthropic factors on native freshwater fish in Brazilian semiarid region. In: DEMPSEY, S. P. (Ed.) **Fish Ecology.** Nova Science Publishers, Inc. New York: USA, 2011. chap 6, p. 115-130.

DE ROSA, L. V. S.; FORESTI, F.; MARTINS, C.; OLIVEIRA, C.; SOBRINHO, P. E.; WASKO, A. P. Cytogenetic analyses of two Curimatidae species (Pisces; Characiformes) from the Paranapanema and Tietê Rivers. **Braz. J. Biol.**, v. 67, n. 2, p. 333-338, 2007.

DELARIVA, R. L.; AGOSTINHO, A. A. Relationship between morphology and diets of six neotropical loricariids. **Journal of Fish Biology**, v. 58, p. 832-847, 2001.

DOOSEY, M. H.; BART JR., H. L. Morphological Variation of the Palatal Organ and Chewing Pad of Catostomidae (Teleostei: Cypriniformes). **Journal of Morphology**, v. 272, p. 1092-1108, 2011.

DOUPÉ, R. G.; KNOTT, M. J. Rapid digestion of fish prey by the highly invasive 'detritivore' *Oreochromis mossambicus*. **Journal of Fish Biology**, v. 76, p. 10191024, 2010.

DRAKE, P., ARIAS, A. M., GALLEGO, L. Biologia de los mugilidos (Osteichthyes, Mugilidae) em los esteros de las salinas de San Fernando (Cadiz) III. Feeding habits and their relationship with digestive tract morphometry. **Inv. Esq.**, v. 48, n. 2, p. 337-367, 1984.

EIRAS-STOFELLA, D. R.; CHARVET-ALMEIDA, P. Ultrastructure (SEM) of the gills of prochilodus scrofa steindachner, 1881 (Pisces, Teleostei). Revista Brasileira de Zoologia, v. 15, p. 279-287, 1998.

ELSHEIKH, E. H. Scanning electron microscopic studies of gill arches and rakers in relation to feeding habits of some fresh water fishes. **The Journal of Basic & Applied Zoology**, v. 66, p. 121-130, 2013.

EMPARN - Agricultural Research Company of Rio Grande do Norte. **Analysis of Accumulated Rainfall (Quantiles) - Year: 2015.** Available at: <http://186.250.20.84/monitoramento/2015/mapas/mapaRN.htm>. Accessed on: 24 Feb. 2016.

FAO - Food and Agriculture Organization of the United Nations. **The State of World Fisheries and Aquaculture.** Rome, 2014. 223 p.

FARBER, S. A.; PACK, M.; HO, S. Y.; JOHNSON, I. D.; WAGNER, D. S.; DOSCH, R.; MULLINS, M. C.; HENDRICKSON, H. S.; HENDRICKSON, E. K.; HALPERN, M. E. Genetic analysis of digestive physiology using fluorescent phospholipid reporters. **Science**, v. 292, p. 1385-1388, 2001.

FAVERO, J. M.; POMPEU, P. S.; VALLADARES, A. C. P. Reproductive aspects of two cichlid species in the Amana Sustainable Development Reserve, Amazonas, Brazil. **Revista Brasileira de Zoociências**, v. 12, n. 2, p. 117-124, 2010.

FEMURN - Federation of Municipalities of Rio Grande do Norte. **Distances of the Municipalities of Rio Grande do Norte to Natal-RN**. Available at: <http://www.femurn.org.br/distancias-de-natal-rn>. Accessed on: 24 Feb. 2016.

FINGER, T. E. Sorting food from stones: the vagal taste system in Goldfish, *Carassius auratus*. **J Comp Physiol A**, v. 194, n. 135-143, 2008.

FISCHER, L. G.; PEREIRA, L. E. D.; VIEIRA, J. P. **Peixes estuarinos e costeiros**. 2. ed. Rio Grande: Luciano Gomes Fischer, 2011. 131 p.

FISHBASE. ***Steindachnerina notonota* (Miranda-Ribeiro, 1937).** Available:< http://www.discoverlife.org/20/q?search=Steindachnerina+notonota&b=FB53123>.

Accessed on: 24 Feb. 2016.

FRANCO, C. R. C. **Ultrastructural study of the stomach of two teleosts: *Hypostomus commersonnii* (REIS et.al; 1990) and *Rhamdia branneri* (HASEMANN, 1911).** 65 f. 1994. Dissertation (Master's Degree in Biological Sciences) - Federal University of Paranà, Curitiba/PR, 1994.

FROESE, R.; PAULY, D. (Eds.). **FishBase** - version 01/2016. Available at: <http://www.fishbase.se/search.php?lang=Portuguese>. Accessed on: Feb. 24, 2016.

FUGI, R. **Feeding strategies used by five species of bottom-feeding fish from the Upper Paranà River/PR-MS.** 142 f. 1993. Dissertation (Master's Degree in Ecology and Natural Resources) - Federal University of Sao Carlos, Sao Carlos/SP, 1993.

FUGI, R.; N.S. HAHN. Food spectrum and morphological relationships with the digestive tract of three species of bottom-feeding fish from the Paranà River, Brazil. **Rev. Bras. Biol.**, Rio de Janeiro, v. 51, n. 4, p. 873-879, 1991.

GERKING, S. D. **Feeding Ecology of Fish.** San Diego: Academic Press, 1994. 416 p.

GERMAN, D. P.; NEUBERGER, D. T.; CALLAHAN, M. N.; LIZARDO, N.R.; EVANS, D. H. Feast to famine: The effects of food quality and quantity on the gut structure and function of a detritivorous catfish (Teleostei: Loricariidae). **Comparative Biochemistry and Physiology**, v. 155, p. 281-293, 2010.

GIORA, J.; FIALHO, C. B. Feeding Biology of *Steindachnerina brevipinna* (Characiformes, Curimatidae), from the Ibicui-Mirim River, Rio Grande do Norte, Brazil. **Iheringia, Sér. Zool.**, v. 93, n. 3, p. 277-281, 2003.

GODINHO, M. H; TOKUMARU, M.; FERRI, A. G. Histology of the digestive tract of *Pimelodus maculatus* Lacepede (PISCES, SILUROIDEI). **Rev. Brasil. Biol.**, v. 30, n. 4, p. 583-593, 1970.

GOOGLE MAPS/EARTH. **5°37'47"S 35°37'9"W**. Available at: <https://www.google.com.br/maps/place/5%C2%B037'47.0%22S+35%C2%B037'09 . 0%22W/@-5.6252999,-35.6083409,2889m/data=!3m1!1e3!4m2!3m1!1s0x0:0x0>. Accessed on: Feb. 24, 2016.

GUISANDE, C.; PELAYO-VILLAMIL, P.; VERA, M.; MANJARRÉS-HERNANDEZ, A.; CARVALHO, M. R.; VARI, R. P.; JIMÉNEZ, L. F.; FERNANDEZ, C.; MARTiNEZ, P.; PRIETO-PIRAQUIVE, E.; GRANADO-LORENCIO, C.; DUQUE, S. R. Ecological Factors and Diversification among Neotropical Characiforms. **International Journal of Ecology**, v. 2012, p. 1-20, 2012.

GURGEL, L. L.; VERANI, J. R.; CHELLAPPA, S. Reproductive Ecology of Prochilodus brevis an Endemic Fish from the Semiarid Region of Brazil. **The Scientific World Journal**, v. 2012, p. 1-7, 2012.

HAHN, N. S.; CUNHA, F. Feeding and Trophic Ecomorphology of *Satanoperca pappaterra* (Pisces, Cichlidae) in the Manso Reservoir, Mato Grosso State, Brazil. **Brazilian Archives of Biology and Technology**, v. 48, n. 6, p. 1007-1012, 2005.

HANSEN, A.; GHOSAL, R.; CAPRIO, J.; CLAUS, A. W.; SORENSEN, P. W. Anatomical and physiological studies of bigheaded carps demonstrate that the epibranchial organ functions as a pharyngeal taste organ. **The Journal of Experimental Biology**, v. 217, p. 3945-3954, 2014.

HARDER, W. The intestine as a diagnostic character in identifying certain Clupeioids (Engraulidae, Clupeidae, Dussumieriidae) and as a morphometric character for comparing anchoveta (*Centengraulis mysticetus*) populations. **InterAme. Trop. Tuna Comm.**, v. 2, n. 8, p. 367-380, 1958.

HEIDMAN, M. K.; HOLLEY, L. L.; CHAMBERS, R. M.; SANDERSON, S. L. Selective feeding on nutrient-rich particles by gizzard shad *Dorosoma cepedianum* does not involve mechanical sorting. **Aquat Biol**, v. 17, p. 129-139, 2012.

HELFMAN, G. S., COLLETTE, B. B.; FACEY, D. E.; BOWEN, B. W. **The diversity of fishes** - Biology, Evolution, and Ecology. USA: Wiley-Blackwell, 2009.

HERNANDEZ, D. R.; PÉREZ GIANESELLI, M.; DOMITROVIC, H. A. Morphology, histology and histochemistry of the digestive system of South American catfish (*Rhamdia quelen*). **Int. J. Morphol.**, v. 27, n. 1, p. 105-111, 2009.

HOLLEY, L. L.; HEIDMAN, M. K.; CHAMBERS, R. M.; SANDERSON, S. L. Mucous contribution to gut nutrient content in American gizzard shad *Dorosoma cepedianum*. **Journal of Fish Biology**, v. 86, p. 1457-1470, 2015.

HONDA, E. M. S. Feeding and Reproduction of *Pseudocurimata gilberti* (QUOY & GAIMARD, 1824) from the Cachoeira River. Master's dissertation. UFPR, Curitiba/PR, 1979. 89p.

HORN, M. H.; CORREA, S. B.; PAROLIN, P.; POLLUX, B. J. A.; ANDERSON, J. T.; LUCAS, C.; WIDMANN, P.; TJIU, A.; GALETTI, M.; GOULDING, M. Seed dispersal by fishes in tropical and temperate fresh waters: The growing evidence. **Acta Oecologica**, v. 37, p. 561-577, 2011.

HOWES, G. S. Review of the genus Brycon (Teleostei; Characoidei). **Bull. Br. Mus. Nat. Hist.**, v. 43, n. 1, p. 1-47, 1982.

HUITING, L. N.; LAROCHE, F. J. F.; FENG, H. The Zebrafish as a Tool to Cancer Drug Discovery. **Austin J Pharmacol Ther**, v. 3, n. 2, p. 1-9, 2015.

IBGE - Brazilian Institute of Geography and Statistics. **Brazilian territorial division -2014.** Available at:

<http://www.ibge.gov.br/home/geociencias/cartografia/default_dtb_int.shtm>.

Accessed on: 24 Feb 2016.

IDEMA - Institute for Sustainable Development and the Environment of Rio Grande do Norte. Programa Estadual de Monitoramento e Fiscalizaçao Ambiental Aèreo - PEMFAA - **Relatório de Sobrevôo: Rio Cearâ-Mirim (executed on 09-102009)**.

Available at:<
http://adcon.rn.gov.br/ACERVO/idema/DOC/DOC000000000005942.PDF>.
Accessed on: 24 Feb. 2016.

IGARN - Rio Grande do Norte State Water Management Institute. **Cearà Mirim Basin** - 07/05/2014. Available at:
<http://adcon.rn.gov.br/ACERVO/IGARN/DOC/DOC000000000028882.PDF>.
Accessed on: 24 Feb. 2016.

JESUS, T. B.; CARVALHO, C. E. V. Use of bioindicators in fish as a tool for assessing environmental contamination by mercury (Hg). **Oecol. Bras.**, v. 12, n. 4, p. 680-693, 2008.

KARDASH, E. Current methods in research using zebrafish. **Mater Methods**, v. 2, n. 109, 2012. Available at:
<http://www.labome.com.br/method/Current-Methods-in-Zebrafish-Research.html>.
Accessed on: 24 Feb. 2016.

KHOJASTEH, S. M. B. The morphology of the post-gastric alimentary canal in teleost fishes: a brief review. **Int. J. Aqu. Sci**, v. 3, n. 2, p. 71-88, 2012.

LAGLER, K. F.; BARDACH, J. E.; MILLER, R. R.; PASSINO, D. R. M.
Ichthyology. 2. Ed. New York: Wiley & Sons, 1977.

LALL, S. P.; TIBBETTS, S. M. Nutrition, Feeding, and Behavior of Fish. **Vet Clin Exot Anim**, v. 12, p. 361-372, 2009.

LEAL, I. R.; SILVA, J. M. C.; TABARELLI, M.; LACHER JR., T. E. Changing the course of biodiversity conservation in the Caatinga of Northeastern Brazil. **Megadiversity**, v. 1, n. 1, p.139-146, 2005.

LEÂO, T. C. C.; ALMEIDA, W. R.; DECHOUM, M.; ZILLER, S. R. **Invasive Alien Species in Northeastern Brazil**: Context, Management and Policies Pùblicas. Recife: Cepan, 2011. 99 p.

LÉVÊQUE, C.; OBERDORFF, T.; PAUGY, D.; STIASSNY, M. L. J.; TEDESCO, P.

A. Global diversity of fish (Pisces) in freshwater. **Hydrobiologia**, v. 595, p. 545567, 2008.

LIMA, F. B.; BRACCINI, M. C.; DiAZ, A. O.; PINHEIRO JÛNIOR, C.; GUIMARÂES, A. C. G. Morphology of the gills *of Steindachnerina brevipinna* (Eigenmann & Eigenmann, 1889) (Characiformes, Curimatidae). Revista Biotemas, v. 22, n.1, p. 87-92, 2009.

LUCINDA, P. H. F.; VARI, R. P. New *Steindachnerina* Species (Teleostei: Characiformes: Curimatidae) from the Rio Tocantins Drainage. **Copeia**, n. 1, p. 142147, 2009.

LUJAN, N. K.; WINEMILLER, K.O.; ARMBRUSTE, J. W. R. Trophic diversity in the evolution and community assembly of loricariid catfishes. **BMC Evolutionary Biology**, v. 12, n. 124, p. 1-12, 2012.

MACRAE, C. A.; PETERSON, R. T. Zebrafish as tools for drug discovery. **Nat Rev Drug Discov**, v. 14, n. 10, p. 721-731, 2015.

MAIA, V. **Técnica Histológica**. 2ª Ed. Sao Paulo: Atheneu, 1979. p. 70-136.

MAKINO, L. C. **Structure, ultrastructure and histochemistry of the digestive tract of *Prochilodus lineatus*. Analysis of the diversity of the intestinal microbiota of**

***Prochilodus lineatus* and *Pterygoplichthys anisitsi*.** 92 f. 2010. Thesis (Doctorate in Aquaculture) - Universidade Estadual Paulista, Jaboticabal, Sao Paulo/SP, 2010.

MALABARBA, L. R.; MALABARBA, M. C. S. L. Phylogeny and classification of neotropical fishes. In: BALDISSEROTTO, B.; CYRINO, J. E. P.; URBINATI, E. C. (Ed.). **Biology and physiology of neotropical freshwater fish**. Jaboticabal: FUNEP; UNESP, 2014. chap. 1, p. 1-12.

MALABARBA, L. R.; MALABARBA, M. C. S. L. Ichthyofauna of the Southern Region.

Ciência & Ambiente, v. 35, p. 55-64, 2007.

MARTIN, P. G.; OLIVEIRA, A. V.; PRIOLI, A. J.; PRIOLI, S. M. A. P. Molecular characterization of *Steindachnerina* (PISCES, CHARACIFORMES) populations from the floodplain of the upper Paranâ River. **CESUMAR**, v. 11, n. 1, p. 7-13, 2009.

MENIN, E. **Comparative Functional Anatomy and Histology of the Digestive System of Six Freshwater Teleostei (Pisces).** 1988. 539 f. Thesis (Doctorate in Sciences) - University of Sao Paulo, Sao Paulo, 1988.

MENIN, E.; MIMURA, O. M. Anatomical aspects of the epibranchial organs of *Prochilodus marggravii* (Walbaum, 1792) and *Prochilodus affinis* (Reinhardt, 1874) (Characiformes, Prochilodontidae). **Revista Ceres**, v. 38, n. 217, p. 229-239, 1991

MILLER, R. V. The morphology and function of the pharyngeal organs in the clupeid, *Dorosoma petenense* (Gunther). Chesapeake Science, v. 5, n. 4, p. 194-199, 1964.

MÒNACO, I. A.; RESENDE, E. K. **Feeding of the detritivorous fish of Tuiuiù Bay, Paraguay River, Pantanal of Mato Grosso do Sul, Brazil.** Research and Development Bulletin, Corumbà: Embrapa Pantanal, 2012. 18 p.

MONTENEGRO, A. K. A.; CRISPIM, M. C.; TORELLI, J.; MARINHO, R. S. A. The influence of the methodology in the results of food diet analysis of *Steindachnerina notonota* (MIRANDA RIBEIRO, 1937), in Taperoà II Dam, Paraiba State. **Revista de Biologia e Ciências da Terra**, n. 1, p. 180-191, 2006.

MONTENEGRO, A. K. A. Biecology of the ichthyofauna of the Taperoà II reservoir, Semi-arid Paraiba, Brazil. 2007. 143 f. Dissertation (Master's Degree in Biological Sciences) - Federal University of Paraiba, Joao Pessoa, 2007.

MONTENEGRO, A. K. A.; TORELLI, J. E. R.; CRISPIM, M. C.; MEDEIROS, A. M. A. Population and feeding structure of *Steindachnerina notonota* Miranda- Ribeiro, 1937 (Actinopterygii, Characiformes, Curimatidae) in Taperoà II dam, semi-arid region of Paraiba, Brazil. **Acta Limnologica Brasiliensia**, v. 23, n. 3, p. 233244, 2011.

MORAES, M. F. P. G.; BARBOLA, I. F.; GUEDES, E. A. C. Feeding and morphological relationships with the digestive system of the "Curimbatâ", *Prochilodus lineatus* (Valenciennes) (Osteichthyes, Prochilodontidae), from a lagoon in southern Brazil. **Revta bras. Zool.**, v. 14, n. 1, p. 169 -180, 1997.

MOREIRA, C. R. **Phylogenetic relationships in the order Characiformes (Teleostei: Ostariophysi).** 468 p. 2007. Thesis (Doctorate in Sciences) - University of Sao Paulo, Sao Paulo/SP, 2007.

MORRISON, C. M.; WRIGHT JR, J. R. A study of the histology of the digestive tract of the Nile tilapia. **Journal of Fish Biology**, v. 54, p. 597-606, 1999.

MUNDAHL, N. D., WISSING, T. E. Selection and digestive efficiencies of gizzard shad feeding on natural detritus and two laboratory diets. **Trans Am Fish Soc**, v. 117, p. 480-487, 1988.

NASCIMENTO, W. S.; BARROS, N. H. C.; ARAÙJO, A. S.; GURGEL, L. L.; CANAN, B.; MOLINA, W. F.; ROSA, R. S.; CHELLAPPA, S. Composition of the ichthyofauna of the river basins of Rio Grande do Norte, Brazil. **Biota Amazônia**, v. 4, n. 1, p. 126-131, 2014.

NELSON, G. J. Epibranchial organs in lower teleostean fishes. **J. Zool.**, v. 153, p. 71-89, 1967.

NELSON, J. A. Breaking wind to survive: fishes that breathe air with their gut. **J Fish Biol.**, v. 84, n. 3, p. 554-576, 2014.

NETTO-FERREIRA, A. L.; VARI, R. P. New Species of *Steindachnerina* (Characiformes: Curimatidae) from the Rio Tapajós, Brazil, and Review of the Genus in the Rio Tapajós and Rio Xingu Basins. **Copeia**, n. 4, p. 523-529, 2011.

OLAOSEBIKAN, B. D.; NZEH, C. Anatomical features and histology of the digestive tract of two African detritivorous Cyprinids, *Labeo senegalensis* (Valenciennes, 1842) and *L. coubie* (Ruppell, 1832) - Abstract. **Nigerian Journal of Fisheries**, v. 5, n. 2, 2008.

OLIVEIRA, A.V.; PRIOLI, A. J.; PRIOLI, S. M. A. P.; RAVANELLI, C. S.; JÛLIO JR., H. F.; PANARARI, R.S. Diversity and genetic distance in populations of *Steindachnerina* in the upper Paranà river floodplain of Brazil. **Genetica**, v. 115, p. 259-267, 2002.

OLIVEIRA, D. C.; BENNEMANN, S. T. Ichthyofauna, food resources and relationships with anthropogenic interference in an urban stream in southern Brazil. **Biota Neotropica**, v. 5, n. 1, p. 95-107, 2005.

OLIVEIRA, F. A.; REIS, L. P.; SOTO-BLANCO, B.; MELO, M. M. Pesticides residues in the *Prochilodus costatus* (Valenciennes, 1850) fish caught in the Sao Francisco River, Brazil. **J Environ Sci Health B**, v. 50, n. 6, p. 398-405, 2015.

OLIVEIRA, J. C. S.; ISAAC, V. J. Diet breadth and niche overlap between *Hypostomus plecostomus* (Linnaeus, 1758) and *Hypostomus emarginatus* (Valenciennes, 1840) (Siluriformes) in the Coaracy Nunes hydroelectric reservoir, Ferreira Gomes, Amapâ-Brazil. **Biota Amazônia**, v. 3, n. 2, p. 116-125, 2013.

OLSSON, C. Autonomic innervation of the fish gut. **Acta histochemica**, v. 111, p. 185-195, 2009.

OLSSON, C. The enteric nervous system. In: GROSELL, M.; FARRELL, A. P.; BRAUNAER, C. J. (Ed.) **Fish physiology**: the multifunctional gut of fish. San Diego: Academic Press, 2011. chap. 8, p. 319-349.

PASLEAU, F.; R. DIOGO; M. CHARDON. The epibranchial organ and its anatomical environment in the Gonorynchiformes, with functional discussions. In: Grande, T.; F. Poyato-Ariza; R. Diogo (Eds.). **Gonorynchiformes and ostariophysan relationships** - a comprehensive review. Oxford, UK: Science Publishers, Taylor & Francis, 2009. p. 140-166.

PEREIRA, R. A. C.; RESENDE, E. K. **Peixes detritivores da planicie inundâvel do rio Miranda, Mato Grosso do Sul, Pantanal.** Corumbà, MS: EMBRAPA-CPAP, 1998. 97 p.

PEREIRA, T. N. A. **Comparative Encephalic Anatomy of Characiformes (Teleostei: Ostariophysi).** 279 f. 2014. Thesis (Doctorate in Sciences) -

University of Sao Paulo, Ribeirao Preto/SP, 2014.

PERESSIN, A.; GONÇALVES, C. S.; BRAGA, F. M. S. Reproductive strategies of two Curimatidae species in a Mogi Guaçu impoundment, upper Paranà River basin, Sao Paulo, Brazil. **Neotropical Ichthyology**, v. 10, n. 4, p. 847-854, 2012.

PESSOA, E. K. R.; SILVA, N. B.; CHELLAPPA, N.T.; SOUZA, A. A.; CHELLAPPA, S. Comparative morphology of the digestive tract of the fish *Hoplias malabaricus* and *Hypostomus pusarum* from the Marechal Dutra reservoir, Rio Grande do Norte, Brazil. **Biota Amazônia**, v. 3, n. 1, p. 48-57, 2013.

QUEIROZ, L. P.; RAPINI, A.; GIULIETTI, A. M. (Ed.). **Towards greater knowledge of the Brazilian Semi-arid Biodiversity.** Brasilia: Ministry of Science and Technology, 2006. 138 p.

RAKERS, S.; GEBERT, M.; UPPALAPATI, S.; MEYER, W.; MADERSON, P.; SELL, A. F.; KRUSE, C.; PAUS, R. "Fish matters': the relevance of fish skin biology to investigative dermatology. **Experimental Dermatology**, v. 19, p. 313-324, 2010.

RAMIRES, M.; CLAUZET, M.; BARRELLA, W.; ROTUNDO, M. M.; SILVANO, R. A. M.; BEGOSSI, A. Fishers' knowledge about fish trophic interactions in the southeastern Brazilian coast. **Journal of Ethnobiology and Ethnomedicine**, v. 11, n. 19, p. 1-11, 2015.

RAMOS, T. P. A.; BARROS-NETO, L. F.; BRITSKI, H. A.; LIMA, S. M. Q. *Parotocinclus seridoensis*, a new hypoptopomatine catfish (Siluriformes:

Loricariidae) from the upper rio Piranhas-Açu basin, northeastern Brazil. **Neotropical Ichthyology**, v. 11, n. 4, p. 787-796, 2013.

RIBEIRO, C. A. O.; FANTA. E. Microscopic Morphology and histochemistry of the digestive system of a tropical freshwater fish *Trichomycterus brasiliensis* (Lütken)(Siluroidei, Trichomycteridae). **Rev. bras. Zool.**, v. 17, n. 4, p. 953-971, 2000.

RICHARDSON, R.; SLANCHEV, K.; KRAUS, C.; KNYPHAUSEN, P.; EMING, S.; HAMMERSCHMIDT, M. Adult zebrafish as a model system for cutaneous wound-healing research. **J Invest Dermatol**, v. 133, n. 6, p. 1655-1665, 2013.

RODRIGUES, S. S.; MENIN, E. Anatomical adaptations of the buccopharyngeal cavity of *Pseudoplatystoma corruscans* (Spix and Agassiz, 1829) (Siluriformes, Pimelodidae) in relation to its feeding habits. **Revista Ceres**, v. 53, n. 305, p. 135-146, Jan.-feb., 2006a.

RODRIGUES, S. S.; MENIN, E. Anatomy of the buccopharyngeal cavity of *Salminus brasiliensis* (Cuvier, 1817) (Pisces, Characidae, Salmininae). **Biotemas**, v. 19, n. 1, p. 41-50, Mar., 2006b.

RODRIGUES, S. S.; MENIN, E. Anatomy of the buccopharyngeal cavity *of Conorhynchos conirostris* (VALENCIENNES, 1840) (SILURIFORMES). **Revista Ceres**, v. 52, n. 304, p. 843-862, 2005.

RODRIGUES, S. S.; NAVARRO, R. D.; MENIN, E. Anatomical adaptations of the buccopharyngeal cavity of *Leporinus macrocephalus* Garavello & Britski, 1988 (Pisces, Characiformes, Anostomidae) in relation to feeding habits. **Biotemas**, v. 19, n. 1, p. 51-58, Mar., 2006.

ROMER, A. S.; PARSONS, T. S. **Comparative anatomy of vertebrates**. Sao Paulo: Atheneu, 1985. 559 p.

ROSA, R. S.; MENEZES, N. A.; BRITSKI, H. A.; COSTA, W. J. E. M.; GROTH, F.. Diversity, distribution patterns and conservation of fish from the Caatinga. In: LEAL, I. R.; TABARELLI, M.; SILVA, J. M. C. **Ecologia e conservaçâo da Caatinga.** Recife: Editora Universitària, Universidade Federal de Pernambuco, 2003. ch. 3, p. 135-180.

ROTTA, M. A. **Aspectos gerais da fisiologia e estrutura do sistema digestivo dos peixes relacionados à piscicultura.** Corumbâ: Embrapa Pantanal, 2003. 48 p.

SALVADOR-JR, L. F.; SALVADOR, G. N.; SANTOS, G. B. Morphology of the digestive tract and feeding habits of *Loricaria lentiginosa* Isbrücker, 1979 in a Brazilian reservoir. **Acta Zoologica**, v. 90, p. 101-109, 2009.

SAMPAIO, A. L. A.; GOULART, E. Neotropical cichlids: trophic ecomorphology. **Oecologia Australis**, v. 15, n. 4, p. 775-798, 2011.

SAMPAIO, T. R.; GRAVENA, W.; GOUVEIA, J. G.; GIULIANO-CAETANO, L.; DIAS, A. L. B microchromosomes in the family Curimatidae (Characiformes): mitotic and meiotic behavior. **CompCytogenet**, v. 5, n. 4, p. 301-13, 2011.

SAMPAIO, T. R.; PIRES, L. B.; VENTURELLI, N. B.; USSO, M. C.; ROSA, R.; DIAS, A. L. Evolutionary trends in the family Curimatidae (Characiformes): inferences from chromosome banding. **CompCytogen**, v. 10, n.1, p. 77-95, 2016.

SAMPAIO, W. M. S.; BELEI, F.; GIONGO, P.; SILVA, W. L. Ichthyofauna, Uberabinha River (Upper Paranaiba River Basin), Triangle Mineiro region, Uberlândia, Minas Gerais, Brazil. **Check List**, v. 8, n. 6, p. 1085-1088, 2012.

SANTOS, L. V. R.; FORESTI, F.; WASKO, A. P.; OLIVEIRA, C.; MARTINS, C. Nucleotide sequence, genomic organization and chromosome localization of 5S rDNA in two species of Curimatidae (Teleostei, Characiformes). **Genetics and Molecular Biology**, v. 29, n. 2, p. 251-256, 2006.

SANTOS, M. L.; ARANTES, F. P.; SANTIAGO, K. B.; SANTOS, J. E. Morphological characteristics of the digestive tract of *Schizodon knerii* (Steindachner, 1875), (Characiformes: Anostomidae): An anatomical, histological and histochemical study. **An Acad Bras Cienc**, v. 87, n. 2, p. 867-878, 2015.

SCHUINGUES, C. O.; LIMA, M. G.; LIMA, A. R.; MARTINS, D. S.; COSTA, G. M. Anatomy of the buccopharyngeal cavity of *Sorubim trigonocephalus* (Siluriformes, Osteichthyes). **Pesq. Vet. Bras.**, v. 33, n. 10, p. 1256-1262, 2013.

SEIXAS FILHO, J. T.; BRAS, J. M.; GOMIDE, A. T. M.; OLIVEIRA, M. G. A.; DONZELE, J. L.; MENIN, E. Functional anatomy and morphometry of the intestine in the freshwater surubim Teleostei (Pisces) (*Pseudoplatystoma coruscans* - Agassiz, 1829). **Revista Brasileira de Zootecnia**, v.30, n.6, p.1670-1680, 2001.

SILVA FILHO, E. G.; SANTANA, F. M. S.; SEVERI, W. Ichthyofauna of the Duas Unas reservoir, Jaboatao river basin, Pernambuco: preliminary results on assemblage composition and structure. **Revista Brasileira de Ciências Agrârias**, v.6, n.2, p.351-361, 2011.

SILVA, M. R.; NATALI, M. R. M.; HAHN, N. S. Histology of the digestive tract of *Satanoperca pappaterra* (Osteichthyes, Cichlidae). **Acta Scientiarum**, v. 34, n. 3, p. 319-326, 2012.

SILVA, N. B. **Morphohistological study of the digestive tract of *Steindachnerina notonota* Ribeiro, 1937 (Characiformes, Curimatidae) and *Hoplias malabaricus* Bloch, 1794 (Characiformes, Erythrinidae) from the Cearà Mirim river in Umari, municipality of Taipu - RN.** 2004. 68 f. Dissertation (Master's Degree in Ecology) - Federal University of Rio Grande do Norte, Natal/RN, 2004.

SILVA, N. B.; GURGEL, H. C. B.; SANTANA, M. D. Histology of the digestive system of sage-grouse, *Steindachnerina notonota* (Miranda Ribeiro, 1937) (Pisces, Curimatidae), from the Cearà Mirim River, Rio Grande do Norte, Brazil. **Boletim do Instituto de Pesca**, Sao Paulo, v. 31, n.1, p. 1-8, 2005.

SILVA-CAMACHO, D. S.; SANTOS, J. N. S.; GOMES, R. S.; ARAÙJO, F. G. Ecomorphological relationships among four Characiformes fish species in a tropical reservoir in South-eastern Brazil. **Zoologia**, v. 31, n. 1, p. 28-34, 2014.

SIMONATO, J. D.; GUEDES, C. L. B.; MARTINEZ, C. B. R. Biochemical, physiological, and histological changes in the neotropical fish Prochilodus lineatus exposed to diesel oil. **Ecotoxicol Environ Saf**, v. 69, n.1, p.112-120, 2008.

STARLING, F.; LAZZARO, X.; CAVALCANTI, C.; MOREIRA, R. Contribution of omnivorous tilapia to eutrophication of a shallow tropical reservoir: evidence from a fish kill. **Freshwater Biology**, v. 47, n. 12, p. 2443-2452, 2002.

STEVENS, C. E.; HUME, I. D. **Comparative physiology of the vertebrate digestive system.** 2nd ed. Cambridge (UK): Cambridge University Press, 2004.

STOSKOPF, M. K. Anatomy. In: **Fish Medicine.** Ed. W. B, Mexico: Saunders Company. 1993. p. 03-31.

TEIXEIRA, I.; BENNEMANN, S. T. Ecomorphology reflecting fish diet in a reservoir in southern Brazil. **Biota Neotropica**, v. 7, n. 2, p. 67-76, 2007.

TORRES, R. A.; MATOSO, D. A.; ARTONI, R. F. Genetics of neotropical fish.

II. Molecular biology of neotropical fish. **Publ. UEPG Ci. Biol. Saùde**, Ponta Grossa, v. 10, n. 2, p. 27-37, jun. 2004.

VARI, R. P. Systematics of the Neotropical characiform genus Steindachnerina Fowler (Pisces, Ostariophysi). **Smithsonian Contributions to Zoology**, v. 507, p.1118, 1991.

VARI, R. P.; MALABARBA, L. R. **Neotropical ichthyology: an overview**. In: MALABARBA, L. R.; REIS, R. E.; VARI, R. P.; LUCENA, Z. M. S.; LUCENA, C. A. S. (Ed.). Phylogeny and classification of Neotropical fish. Porto Alegre: Edipucrs, 1998. p. 1-12.

VARI, R. P.; MALABARBA, L. R. Neotropical Ichthyology: an overview. In: MALABARBA, L. R.; REIS, R. E.; VARI, R. P.; LUCENA, Z. M. S.; LUCENA, C. A. S. **Phylogeny and Classification of Neotropical Fishes**. Porto Alegre: Edipucrs, 1998. p. 1-11.

VENERE, P. C; SOUZA, I. L.; SILVA, L. K.; DOS ANJOS, M. B.; DE OLIVEIRA, R. R.; GALETTI JR, P. M. Recent chromosome diversification in the evolutionary radiation of the freshwater fish family Curimatidae (Characiformes). **Journal of Fish Biology**, v. 72, p. 1976-1989, 2008.

WILSON, J. M.; CASTRO, L. F. C. Morphological diversity of the gastrointestinal tract in fishes. In: GROSELL, M.; FARRELL, A. P.; BRAUNAER, C. J. (Ed.) **Fish physiology**: the multifunctional gut of fish. San Diego: Academic Press, 2011. chap. 1, p. 1-55.

WINEMILLER, K. O.; KELSO-WINEMILLER, L. C. Food habits of tilapiine cichlids of the upper Zambezi river and floodplain during descending phase of the hydrologic cycle. **Journal of Fish Biology**, v. 63, n. 1, p. 120-128, 2003.

YAMADA, S.; YAMAMOTO, K.; IKED, T.; YANAGIGUCHI, K.; HAYASHI, Y. Potency of Fish Collagen as a Scaffold for Regenerative Medicine. **BioMed Research International**, v. 2014, p. 1-8, 2014.

YAMAMOTO, K.; IGAWA, K.; SUGIMOTO, K.; YOSHIZAWA, Y.; YANAGIGUCHI, K.; IKEDA, T.; YAMADA, S.; HAYASHI, Y. Biological Safety of Fish (Tilapia) Collagen. **BioMed Research International**, v. 2014, p. 1-9, 2014.

YOSSA, M. I.; ARAUJO-LIMA, C. A. R. M. Detritivory in two Amazonian fish species. **Journal of Fish Biology**, v. 52, p. 1141-1153, 1998.

ZAGANINI, R. L.; VIDOTTO-MAGNONI, A. P.; CARVALHO, E. D. Ontogenetic diet shifts of *Oreochromis niloticus* and *Tilapia rendalli* of the Barra Bonita reservoir (Tietê river, Sao Paulo State, Brazil). **Acta Scientiarum**, v. 34, n. 3, p. 255-262, 2012.

ZAVALA-CAMIN, L. A. **Introduction to studies on natural feeding in fish.** Maringà: Eduem Nupelia, 1996. 129p.

MIX
Papier aus verantwortungsvollen Quellen
Paper from responsible sources
FSC® C105338

Printed by Books on Demand GmbH, Norderstedt / Germany